JOINT
APPLICATION
DESIGN®

JOINT APPLICATION DESIGN®

How to Design
Quality Systems
in 40% Less Time

JANE WOOD   DENISE SILVER

WILEY

JOHN WILEY & SONS

New York • Chichester • Brisbane • Toronto

Analyst/Designer Toolkit is a registered trademark of Yourdon, Inc.
DCF/PLUS is a registered trademark of Image Sciences.
Design Aid is a registered trademark of the Nastec Corp.
Document Composition Facility (Script/VS) is a registered trademark of the IBM Corp.
Excelerator is a registered trademark of the Index Technology Corp.
Joint Application Design is a registered trademark of IBM Corp.
Information Engineering Workbench is a registered trademark of KnowledgeWare, Inc.
Multimate is a registered trademark of the Multimate International Corp.
pfs:write is a registered trademark of the Software Publishing Corp.
Post-it is a registered trademark of the 3M Corp.
Vis-a-Vis is a registered trademark of the Sanford Corp.
WordPerfect is a registered trademark of the WordPerfect Corp.

Library of Congress Cataloging-in-Publication Data

Wood, Jane, 1946–
    Joint application design : how to design quality systems in 40%
less time / Jane Wood, Denise Silver.
        p.   cm.
    Bibliography: p.
    1. System design—Methodology.   I. Silver, Denise.   II. Title.
    QA76.9.S88W65   1989
    004.2'1—dc20   89-32006
    ISBN 0-471-50462-9   CIP

Printed in the United States of America

10   9   8   7   6   5   4   3   2   1

for Sarah and Allison

PREFACE

This book describes a methodology for designing computer systems. Called Joint Application Design® (or JAD, for short), this methodology centers on a three- to five-day workshop that brings together users and MIS (Management Information Systems) professionals. Under the direction of a leader, these people define user requirements and system specifications. The products of this workshop include definitions of data elements, work flow, screens, and reports.

The advantages of using the JAD methodology include a dramatic shortening of the time it takes to get a business application from user request to the start of programming. Equally important, JAD improves the quality of systems design by focusing on the up-front portion of the development cycle, thus reducing the likelihood of errors that are expensive to correct later on.

In our travels to various conferences, we have met many people who, caught up in the challenges of designing systems, want to know more about JAD. Many of these people have heard about the methodology being used in various organizations. Others have read about it. Some have heard James Martin or other notable software engineers talk about JAD as a productivity tool. Now they want to get past the generalities and down to specifics—how do you actually run a JAD? Consequently, we decided to write a detailed, how-to description of the methodology that answers such questions as:

- What do you need to implement JAD in an organization?
- *Who* does *what* during each step of the process?
- What are the ingredients for a successful JAD?

This book guides you through the five phases of JAD. It offers techniques that you can use not only for designing application systems, but in any situation where you are leading a group of people through a decision-making process. It includes CASE (Computer Assisted Software Engineering) techniques, useful checklists, workplans, sample agendas, and documentation forms that you can copy directly from the book. And it describes the psychology of JAD—how to deal with the kinds of people and situations that can turn a productive meeting into a rambling bull session that accomplishes nothing.

Already you may be skeptical because you have read systems design books before. You have plodded through turgid prose, been slowed by technical jargon, and found you cannot translate an academic methodology into something that works for you.

This book is written for practical use. After reading it, you will know how to conduct all phases of JAD from interviewing users, to running the sessions, to preparing the final document.

As you read this book, consider what the data processing industry is up against—the real world of backlogs, time and budget constraints, maintenance, and endless meetings about more systems to design, construct, and further maintain. JAD will not rid the world of these realities. The 70 billion lines of COBOL code that are maintained today will still be there. But JAD does guarantee that *you will design better systems in a shorter time.* User needs will be clearly identified up-front, the first time. And programs will require less maintenance later on. All this translates into quantifiable savings of time, money, and effort.

Jane Wood
Denise Silver

Philadelphia
June, 1989

ACKNOWLEDGMENTS

For you, the reader, acknowledgments are probably the most uninteresting part of the book. But for us this was the most pleasurable part of the book to write. We are delighted to acknowledge those who helped with the endeavor.

First, we would like to thank the Provident Mutual Life Insurance Company. Certainly, this project would never have gotten off the ground if we had not been working in the kind of creative, professional environment Provident has offered. The company has been progressive enough to bring in a methodology such as JAD, supportive enough to provide the necessary resources to make it work, and broad-minded enough to enable us to write about it. We would especially like to thank Gregg Afflerbach and Guy Edwards.

For his many contributions to the book, we are indebted to Denise's husband, Bruce Schwartz. In the year it took us to write this book, he postponed several home improvement projects to spend time with Allison, his three-year-old daughter, so that Denise could work with her coauthor. Aside from playing "Mr. Mom," Bruce made a direct contribution to the project: he read, edited, and critiqued the entire text. His comprehensive editing ranged from recommendations on overall structure to micro modifications. Bruce is a master of the comma, hyphen, and semicolon. He can spell words like "solely" and "impasse," and he knows the difference between "eluding" and "alluding." His practical (and sometimes picky) suggestions helped to improve the work. For this we are truly grateful.

We deeply appreciate the help of Michael Kowalski for his creative concepts and numerous research contributions. This master of the metaphor suggested several communication techniques that helped bring elusive concepts to life. When challenged with a question, he was able to find the answer by the next day, along with supporting documentation from the most obscure sources.

We are grateful to Joe Darcy (Provident Mutual's most creatively eccentric JAD leader), who has been involved with reworking the JAD methodology since we brought it into the company.

Finally, we would like to thank the proprietors and waitresses of our favorite diners in the Delaware Valley where we spent hours at a time working on this project. These were the Dining Car, Golden Dawn, Club House, Windsor, Diamond, and both Golden Eagles diners. We owe a great deal to the endless cup of coffee.

We are fortunate to have had the help and support of all these people.

CONTENTS

LIST OF ILLUSTRATIONS

PART

1

IS JAD FOR YOU?

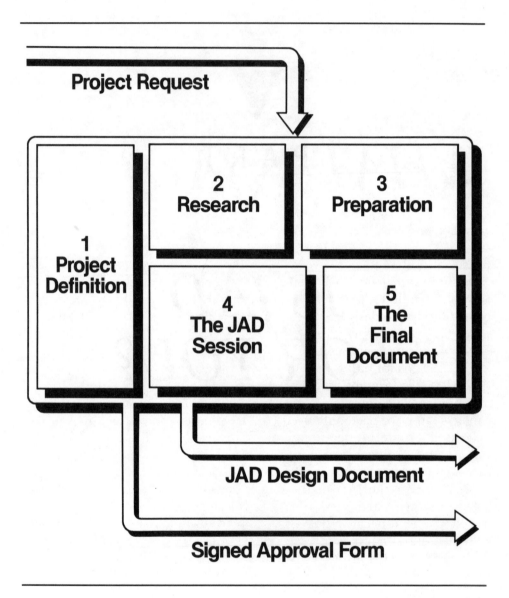

Project Request

1
Project
Definition

2
Research

3
Preparation

4
The JAD
Session

5
The
Final
Document

JAD Design Document

Signed Approval Form

ONE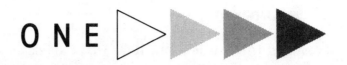

WHAT IS JAD?

JAD is a joint venture between users and data processing professionals; thus the name *Joint Application Design*. It centers around a structured workshop (called a JAD session) where these users and DP professionals come together to design a computer system. It involves a detailed agenda, visual aids, a leader who moderates the session, and a scribe who records the agreed-upon specifications. It culminates in a final document containing definitions for data elements, work flow, screens, and reports.

In 1985, we brought JAD into our organization, the Provident Mutual Life Insurance Company. Three years and more than 30 system design projects later, we have enhanced JAD to include five distinct phases. We have used it to define requirements for insurance applications and financial systems. We have found it especially helpful for any kind of system that involves users from more than one department.

Users prefer the JAD methodology because they become directly involved in the system design. MIS (Management Information Systems) people prefer it because the JAD leader (whose main role is to interface with end users) handles the tasks of coordinating and documenting the design effort. This allows MIS project managers and their staffs more time to concentrate directly on the nuts-and-bolts of system design and programming.

WHERE DID JAD COME FROM?

In 1977, Chuck Morris of IBM conceived some innovative ways of getting users together with MIS to work out plans to install distributed systems.

From this, JAD was born. In 1980, IBM Canada adopted and refined the methodology. Since then, it has been used by large and small companies around the world for all kinds of applications. There are insurance JADs, manufacturing JADs, and utility company JADs. There are JADs for data base design, for estimating the costs of large-scale projects, and for creating Requests for Proposal. There are even JADs to plan other JADs.

There are many methods used to organize and run interactive design sessions. The method described in this book is based on IBM's original JAD.

WHAT CAN JAD DO FOR YOU?

There is nothing mysterious about JAD. It is simple. It is structured. And it saves time. This interactive process leads to success in translating user needs into quality computer systems. It allows you to:

- *Simplify*. JAD consolidates months of design meetings, follow-up meetings, review meetings, telephone-call meetings, and we-just-want-to-clear-this-up meetings into *one* structured workshop.

- *Identify*. The beginning phases of any project are laden with unresolved questions (open issues) that require decisions or further research. JAD offers a way to identify and document these issues as well as to track them until they are resolved. Also, JAD identifies participants, that is, *who* should be involved in designing the application. Having the right people in the workshop (the JAD session) is critical to creating a successful system design.

- *Quantify*. JAD quantifies what is needed for the application. This includes handling questions like: How many characters are in each data element? Which fields are needed on the screen? Where should they be located? And who should receive the reports?

- *Clarify*. Designing systems is an evolutionary process. JAD has built-in ways to track specifications as they migrate through the early phases. However, these constantly changing specifications must be crystallized and clarified at some point in time, then captured and documented. That point in time is the JAD session, and the final document is the synthesis of all user requirements agreed upon in the session.

- *Unify*. JAD ties together every step of the design process. It is logical, structured, and easy to use. What you get from one phase ties directly to the next. For example, the information gathered in the session can be added directly into the final document. No translation, evaluation, contemplation, or whatever-ation is required. The output from one step is the input to the next.

- *Satisfy*. Because of the ongoing user involvement, JAD results in satisfaction. The users designed the system; therefore it is *their* system. Since shared participation brings a share in the outcome, they become committed to the system's success.

Before looking at the five phases of JAD, let's explore how systems are traditionally designed.

HOW SYSTEMS ARE DESIGNED

Henry Ford was not the first to manufacture automobiles. The great innovation that he brought to industry was the assembly line—a method of *organizing* machinery, materials, and labor so that a car could be put together much faster and cheaper than ever before. The JAD methodology does something similar for computer systems design.

The goal in systems design is simply stated: Figure out what the users really need, and then set up a computer system that will do it. Traditional methods for defining these needs have several built-in delay factors that get worse as more people become involved.

Traditional Systems Design

In most organizations, the systems development life cycle begins with the identification of a need, the assignment of a project leader and team, and often, the selection of a catchy acronym for the project. The leader pursues a series of separate meetings with the people who will use the system or be affected by it.

The leader continues these meetings over time. Since the people he or she is meeting with have other work to do, they are not so easy to reach. But eventually, having documented everything possible, the leader translates copious notes into his or her own technical terminology. That's when it becomes apparent that the requirements from, say, Accounting, don't mesh with what the Sales department wants.

The leader calls Sales, and finds out the contact there is in the field and will not be back until tomorrow. Next day the leader reaches Sales, gets the information, calls Accounting, and of course the person in Accounting is now out of the office, and so on.

When the leader finally gets everyone in agreement, alas, he or she discovers that even *more* people should have been consulted because their needs require something entirely different. In the end, everyone is reluctant to "sign off" on the specifications.

This *slow communication* and *long feedback time* is one reason the traditional method is so time consuming. You can see why the communication problem grows worse as more people must be brought into consensus.

Other problems with traditional systems design methods can be described, in broad terms, as *psychological* and *political*. They result in such problems as:

- the *we-versus-they* gap that forms when users and MIS remain sheltered in their own environments. As the gap widens, users begin to view their MIS departments as a kind of electronic priesthood of technical enthusiasts who think they know exactly what the users want without even having to ask. The users might see MIS as an ivory tower of bits and bytes commanding users to "open your mouth, close your eyes, here comes another DP surprise." As James Martin said:

  When the traditional systems analyst and potential end users first meet face to face, they come from widely different cultures. It is rather like a Victorian missionary first entering an African village. (Martin, 1984)

- the *I've-changed-my-mind* gap that happens when meetings are spread over time. What users want this week can be substantially different from what they wanted last week. Such changes are only natural. But you do not want to spend your time documenting these ongoing changes in a series of separate meetings. You need to nail down the specifications all at once with everyone present, thereby getting group consensus at that point in time.

- the *separate-views* gap between everyone involved. Separate meetings breed separate views of what the system should do. People have their own individual ideas of how the system should be designed. On the other hand, when everyone participates in the same meeting, a common view of the system is established.

Systems Design Using JAD

JAD centers around one structured workshop session. Everyone gets together in a room and talks it out. Everyone hears what the rest of the group has to say. There's no delay between question and answer and no "telephone tag" or waiting for memos to come back. This approach brings certain characteristics to the systems design process.

First, there is *commitment*. Participants in the session are under a mandate from management to get the problem solved. Mandates alone cannot guarantee commitment, of course. Rather, the participants' involvement in the design process gives them a vested interest in the final product. This is the source of true commitment. From the start, they are involved in building the application—so they want it to succeed. Everyone needed to make decisions is there in the same room. They define work flow and data elements. They design screens and reports. They see the results on a blackboard in front of them. They work together in an organized forum until everyone agrees, "Yes, that screen works for us." No second rounds. No delays like, "Well, let's run it by Marketing," because Marketing is right there. All of it is done in those three to five days. The participants emerge from the workshop with a product they have all agreed to. The responsibility is shared.

Second, the *physical* and *social* settings of the workshop encourage bonds among participants that make them want to work together. This is achieved by keeping everyone together long enough for them to get to know one another. They are isolated from their normal work environments and, hence, from their "home-grown" ways of doing things. The resulting camaraderie is not unlike the group spirit that develops among people who sit on jury duty, or find themselves lost at sea.

Third, JAD eliminates many of the problems with traditional meetings. As you know, the business world thrives on meetings. Taking a conservative estimate of five hours of meetings per week, managers spend more than 10,000 hours of their lives in meetings. In this country alone, 11 million meetings are held daily. Most companies spend between 7 and 15 percent of their personnel budgets directly on meetings. You cannot entirely eliminate these meetings. (How could you possibly design a system without getting the people together?) But JAD can make these meetings less of a traditional grind. With JAD, these meetings are less frequent, more structured, and more productive. An agenda provides the structure. A leader directs the process. Visual aids clarify concepts being discussed. And the group dynamics, with constant feedback, stimulate creativity.

All in all, JAD channels the participants' efforts along productive lines. As one project manager in our company said:

> The primary benefit of the JAD methodology is that the structure forces the key players to sit eyeball-to-eyeball to discuss the project, present their own viewpoints, answer questions, and arrive at mutually agreeable solutions. All this happens in a relatively short time frame.
>
> —Herb Mullin
> Systems Development Officer
> Provident Mutual Life Insurance Company

THE FIVE PHASES OF JAD

A favorite endeavor of anyone involved in defining a process is to divide it into phases. And so it is with JAD. The methodology evolves from a definite start ("Help, we need a new system!") to a clear finish ("Here are the completed specifications."). Because it is a formal, organized methodology, the phases are logical, distinct, and easy to comprehend.

JAD has five phases. Each is covered in detail in Chapters 3 through 7. The following discussion illustrates these phases using an example in which you have been hired to develop a special-purpose robot. The robot's purpose is to prepare and serve meals in the corporate dining room. (Good help is so hard to find these days.) As you review the phases, notice how each one ties into the one that follows.

Phase 1: Project Definition

This is where all projects begin. In the first phase of designing the robot, you (the person leading the JAD) interview the managers from user departments (those who requested the robot) and MIS (those who will build it). These meetings identify the robot's purpose (Will it do cooking, gardening, or back rubs?), scope (Who will use it and how often?), and objectives (How delicious must the food taste?). Functions are identified (Exactly what do you want from the robot—Eggs Benedict every morning in bed, or just a bagel on the run?). You list some basic assumptions (the robot should keep to a low-cholesterol menu) and identify open issues (one person prefers that the menu include the finest Chablis, while the other wants the house wine). What you gather in these meetings is used to create a document called the *Management Definition Guide*.

In this phase, you also identify the JAD team and schedule the session. A successful session depends on having the right people in the room. They must be able to make binding decisions for their departments. In the robot example, all departments directly affected by the robot should be represented. Much of Chapter 3 tells how to select these people.

Phase 2: Research

This phase involves gathering more details about the user requirements. In the robot example, you meet with the actual people who will use it. You familiarize yourself with the existing system. (Do they already have a robot? What problems are they having with it? What enhancements do they need?) Work flow is defined. Proposed definitions are gathered for data elements, screens (this is a screen-driven robot system), and reports.

Design issues are considered (How will the robot benedict the eggs, cacciatore the chicken, and bourguignon the boeuf? And what are all the physical motions required to flip a pancake?). Based on this research, an agenda is prepared listing what needs to be decided in the session.

Phase 3: Preparation

Everything you need for the session is made ready in this phase. You prepare a script, visual aids, and a *Working Document.* The script guides you through the session. Flip charts, magnetics, and overheads, combined with some other interesting techniques, allow you to steer users through defining their requirements. The Working Document, which contains the proposed specifications gathered in phase two, is the basis for what will be covered in the session.

Phase 4: The JAD Session

This is the heart of JAD—the actual workshop session. For three to five days, you guide the team of robot designers in defining user and system requirements that are based on the Working Document. This includes defining work flow, data elements, screens, and reports. Agreed-upon decisions are documented on forms that tie directly to the final document.

Phase 5: The Final Document

The information captured on the scribe forms is used to produce the final *JAD Design Document.* This comprehensive synthesis of robot requirements is compiled and distributed to the JAD team for their review. A review session is held with all the participants. Changes are discussed and noted and key participants sign an approval form. The JAD process is now complete and program specification can begin. Figure 1–1 shows the five JAD phases and their resulting output.

If the development effort does not involve the users in an organized approach such as described here, there is a good chance the final system will not meet their needs. In the case of the robot waiter, you could end up with a mentally unstable block of metal that rakes yards with a vacuum cleaner instead of basting birds with a brush. By proceeding through the JAD phases, however, you will end up with a state-of-the-art twentieth-century roving robot that cooks delicious, delectible, *bon appetit* meals morning, noon, and night and (if it's in the specs) gives you back rubs at the end of its hard day's work.

No.	Phase	Resulting Output
1	Project Definition	Management Definition Guide
2	Research	Work flow Preliminary specifications JAD session agenda
3	Preparation	Working Document JAD session script Overheads, flip charts, magnetics
4	The JAD Session	Completed scribe forms
5	The Final Document	JAD Design Document Signed approval form

Figure 1-1 The JAD phases

WHAT IS "STRUCTURED METHODOLOGY"?

The adjective "structured" seems to permeate the data processing world. We have structured analysis, structured design, and structured programming, all entwined in our structured lives.

Within JAD, structure means this: Information is gathered in a distinct way and passed from phase to phase to produce the final product. The output from one phase is the input to the next. If you like clarity and lack of illusiveness, you will be comfortable with this methodology. There is nothing left to the imagination.

With JAD, tasks do not vary much from person to person or from project to project. The *what* of tasks changes (for example, payroll reports differ from fixed assets reports), but the *how* of tasks remains the same (the reports are designed using the same method).

Every step is an organized procedure. For example, the workshop itself has structure. A predetermined agenda prevents uncontrolled brainstorming sessions that lead nowhere. And let's look at documentation. A designated person (called the *scribe*) is responsible for documenting decisions made in the session. Without an organized approach, the documentation would take on the flavor of that scribe: How does he or she take notes? Did that person have an influential teacher in high school who taught the wonders of outlining? What parts of the current system interest the scribe most? What are his or her feelings about the people making decisions? One person may be an adversary, another a friend. Naturally, the scribe would tend to document the comments of each one quite differently.

JAD counteracts these idiosyncrasies. Specially prepared scribe forms dictate exactly what to document for each situation and a JAD leader controls when to document. Then the information on these forms be-comes a specific part of the final document. The forms mirror the document.

A NOTE ON TERMINOLOGY

Every industry has its jargon. Some rely heavily on acronyms. Sometimes these acronyms become words in themselves. The word "wimp," for example, was revived from the acronym for Weakly Interactive Massive Particles, a contribution from the world of particle physics. Once unleashed into common usage, it came to denote weakness in general ("the wimp factor"). Next it mutated into a verb form ("wimp out"). In the same way, when people work with the JAD methodology on a daily basis, the acronym JAD can start taking on many forms. Here are some examples of how it has come to be used and abused in our organization:

The Word	Part of Speech	Used in a Sentence
JAD	noun	*JAD* has worked well for us.
JAD	adjective	We attended the *JAD* session.
JAD	verb	To *JAD* or not to *JAD* . . .
JADding	present participle	They've been *JADding* all day.
JADded	past participle	That project should definitely be *JADded.*
JADable	adjective	People will want to JAD any-thing that seems *JADable.*

For clarity's sake, we have tried to keep these variations to a minimum. In this book, you will find:

- JAD, the methodology (for example, is JAD for you?)
- JAD, the particular project (for example, the Fixed Assets JAD)
- JAD, the workshop (for example, the JAD session)

USING THIS BOOK

You can use this book to determine if you want to try JAD in your organization. You can use it as an overall planning guide to manage JADs. Or, you can use it as an instruction manual to guide you through each part of the process.

The best way for you to use this book depends on whether you are looking for the details or the overview. The information you are looking for probably falls within two basic categories that ask: Is JAD for you? or How do you run a JAD? There are specific chapters that answer each question.

Is JAD for You?

If you have overview questions like "Will JAD work for us?" or "How do we implement the methodology?", use this book to evaluate what you actually need to do JADs. You do *not* need to know such details as how to set up the room for the session (at least, not yet). You *do*, however, need to know what resources are required to make JADs happen and generally how the methodology works. Therefore, you should read:

- Chapter 2 (When to Use JAD)—describes how JAD fits into the systems development life cycle, how companies are using JAD, and how to set criteria.

- Chapter 3 (Phase 1: Project Definition)—describes what people you need to run a JAD and how to select the JAD team.

- Chapter 9 (Kinds of JADs)—describes how to tailor sessions for different needs. For example, you can customize the methodology for one project (single JAD), for larger projects (multiple JADs), or for special situations, such as JADs to plan other JADs, to determine priorities for many different projects, and to prepare Requests for Proposal (RFPs).

- Chapter 13 (Where to Go from Here)—discusses JAD benefits (to help in your cost justification process), the pilot project, training the JAD leader, and how to measure JAD success.

You can skip Chapters 3 through 7, which describe the JAD phases. Focus less on how the methodology works and more on what you need to make it happen.

How Do You Actually Run a JAD?

On the other hand, you may have already decided to use JAD. You want to know the details of how to make it work. In this case, you can benefit from all the chapters because you need the overview as well as the details. However, most of the information you need is in these chapters:

- Chapters 3 through 7—contain the recipe for running a JAD. They describe the five phases: Project Definition, Research, Preparation, the JAD Session, and the Final Document. These chapters present step-by-step descriptions of what to do for each of the five JAD phases, including:

- how to define data elements, screens, and reports.
- how to determine work flow. A customized approach shows a way to document work flow that you can use before, during, or totally independent of the JAD methodology.
- how to determine the session agenda.
- how to produce a JAD session script to guide you comfortably through the terrain of an actual session.
- how to prepare visual aids.
- how to set up the meeting room. This shows the most effective table arrangement and describes what you need to prepare the room for the session. (It's easy to overlook details such as making sure you have a spare bulb for the overhead projector.)
- how to produce the final document. This includes organizing source documents, using the most time-efficient editing process, as well as tracking distribution.
- samples of which memos to send to whom.

- Chapter 8 (JAD Psychology)—covers the people aspect of JADs, such as handling conflict, chilling the dominator, and encouraging shy users. This information applies to any forum where you need to guide people through a decision-making process.

- Chapter 10 (Tools and Techniques)—offers timesaving and quality-enhancing approaches to use throughout the phases. It covers such things as:

- work plans that can be used not only for JADs but for any project planning.
- CASE tools.
- samples of actual forms used in the session to document decisions. You can copy these forms directly from the book or modify them for special situations.
- checklists that help track various aspects of the JAD process, such as an inventory checklist to determine which supplies to order before the session.
- standard templates.

- Chapter 11 (Case Study: A Real Live JAD)—ties together all the information by describing a typical JAD project. It discusses problems we encountered and how we handled them.

- Chapter 12 (JAD Episodes)—highlights segments from other JAD projects to show more real-world examples.

A Note About "You"

Throughout the book, we refer to *you*. Let's define who this "you" is. You are the one managing the JAD project. You are the JAD leader who handles the pre-session interviews and preparation, the actual three- to five-day session, and the post-session production of the final document. In other words, although you will have people helping you throughout the phases, you are the one responsible for the JAD project from start to finish.

You, the reader, may not actually be this person leading the JADs. Perhaps you are a project manager, a programmer analyst, or the manager who will be hiring the JAD leader. In any case, you will understand the methodology best if you see it through the JAD leader's eyes. Just remember, the "you" in this book is the JAD leader.

**Systems
Development
Life Cycle**

**Setting
JAD
Criteria**

**How Companies
Are Using JAD**

T W O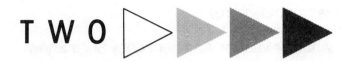

WHEN TO USE JAD

There was a time when hardware was your grandfather's collection of nuts and bolts and bytes were what you took out of peanut butter sandwiches. Now we have online data bases, expert systems, interactive debuggers, code optimizers, code generators, and CASE (Computer Assisted Software Engineering) tools all supported by sophisticated operating systems and complex hardware and telecommunications architectures. We invested in these state-of-the-art tools to keep up with the hardware technology and the backlog of user requests that had burgeoned beyond our expectations.

> According to the findings in a recent survey of over 1,000 organizations, the software application backlog averages 88 months. (McClure, 1988)

As each new productivity tool was installed, we saw new light at the end of the tunnel. But alas, as the adage goes, "Beware of the light at the end of the tunnel, it could be a train coming your way." And thus it was for MIS, a train loaded with mushrooming lines of code and more demanding users. So we have gone in search of more productivity tools to solve our problems.

Another Vendor, Another Fancy Slide Show

Have you attended a product seminar lately? You know, the ones with coffee, muffins, white tablecloths, and friendly vendors with business cards. One thing you can always count on—three slides into the slick presentation, up comes the frame about the systems development life cycle and how we are spending too much time at the end of the cycle instead of the beginning. Pie charts with percentages tell the story. Glib assertions back it up:

> Software maintenance costs have grown to $30 billion per year to support 70 billion lines of COBOL code. (Petzold, 1987)

> Error removal constitutes up to 40% of the cost of a system. Between 45% and 65% of these errors are made in system design. (Rush, 1985)

> Changing a program *after* installation costs 100 times more than changing it in the design phase. (Hennie, 1985)

The vendors elaborate on this dilemma and tell us how we are not dealing with the problem. As the philospher Eric Bush said, "Old computers go into museums. Old programs go into production." Because of tight time constraints, insufficient budgets, and sketchy project requests, we are forced to throw our resources further and deeper into the development cycle. The vendors tell us we should spend 40 percent of our time in up-front analysis instead of an undeterminable (or some may say *interminable*) amount of time in coding and testing. In other words, we need error prevention now or we will face error detection and correction later on.

The vendors offer partial solutions—productivity tools for one segment of the problem. But this does not solve the dilemma they have so well defined. What about tools for overall analysis and design? How do we interview users? How do we translate what they want into actual specifications? How do we handle the contradictory needs among the different departments that will use the system?

JAD provides an answer to these questions. Unlike traditional vendor solutions, JAD is not a software package with an annual maintenance fee or a piece of hardware with a myriad of megabytes and a plethora of peripherals. JAD is a methodology that can be easily learned and practiced.

> Many attempts to improve productivity focus on the programming effort, which is only about 10 percent of the total effort. JAD focuses

on the 90 percent part—the specifications and how the system will operate.

—Guy Edwards
VP, Information Services Department
Provident Mutual Life Insurance Company

WHERE JAD FITS INTO THE SYSTEMS DEVELOPMENT LIFE CYCLE

Systems development life cycles come in as many flavors as ice cream. Some companies use commercial packages such as Method/1,™ SDM/70,™ and Stradis,® while others just get the job done and call it nothing in particular. Furthermore, when it comes to naming the phases within these cycles, the variations are as many as the people who have tried to define them. But no matter how you name the phases in whatever systems development life cycle approach you use, JAD can be part of the process.

Figure 2–1 shows how JAD fits into a typical systems development life cycle. As the chart shows, JAD supports the first three phases of the cycle. It contributes to:

* *Initiation.* This phase defines the problem or opportunity the user has identified. Some companies initiate projects with a feasibility study, others with a project request form or system proposal. JAD can handle all aspects of this phase except cost justification, which

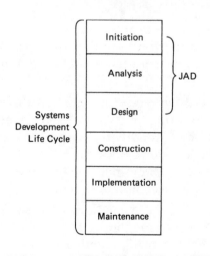

Figure 2-1 JAD in the systems development life cycle

is generally done before a JAD begins. The Management Definition Guide (described in Chapter 3) provides the overview information you need, including the purpose, scope, and objectives of the system.

- *Analysis.* This phase determines what the system will do. It answers such questions as: What functions will the system provide? and What is the work flow?

- *Design.* While the analysis phase defines *what* users need, the design phase determines *how* this will be accomplished. JAD handles design specifications for data elements, screens, and reports. All this information goes into the final document and is passed along to the programming department, which sets up the file structures and writes the code.

After the JAD

After JAD has accomplished its role in the Initiation, Analysis, and Design phases, the systems development life cycle continues along its traditional path. In theory, JAD can be used for other systems development life cycle phases but, in fact, it is rarely used after the design phase. Nevertheless, the people who participated in the JAD session can be useful in later phases. They can help with testing, user documentation, and training.

HOW ARE COMPANIES USING JAD?

The following examples show how companies are using JAD today:

- *New systems design.* An insurance company decides to build an online system that allows the field force to view policy information from their offices across the country. A JAD is held to define the requirements.

- *Modifications to existing systems.* Due to growth at a manufacturing company, its accounting system no longer meets user needs. The company would like to modify the existing system instead of buying a new one. A JAD is held to define the changes to this system. If the company decides the system needs more than just modifications, a JAD could be held for a complete rewrite.

- *Automating manual systems.* A medium-sized company with two small divisions has been growing, but still continues to do its payroll by hand. A JAD is held to bring together the users in the home office and the two divisions to design a system that will automate the manual payroll.

- *Conversions.* A hospital will be converting from one software package

to another. The new package must be tailored to fit into the organization. A JAD is held to define all facets of the conversion.

• *Acquisitions*. An expanding enterprise acquires a subsidiary that will
come under the umbrella of existing software applications in the
parent company. A JAD is held to define the requirements for this
consolidation.

You can see that JAD has many applications. It can be used for almost
any medium- to large-sized project. But how do you determine which
projects should use the JAD methodology? The following describes how to
set criteria.

SETTING JAD CRITERIA

JAD has a way of catching on in the organization. Overnight, skeptical
users turn into strong JAD proponents. People will want to use JAD for
anything that seems "JADable." This sounds good, but it can lead to
misuses of the methodology, especially if you have limited resources. (And
who has unlimited resources?) For example, you might not want to use the
methodology in the following circumstances:

• An MIS project manager and user are having trouble working together and you are called in solely to buffer a personality conflict
between MIS and the user department.
• The actuaries can't seem to find time to sit down and crunch out
those detailed calculations, so a JAD is proposed solely to force a
time commitment to the project.
• An MIS project leader would like to hold a JAD session just to define
program specifications, including field edits for payroll screens and
detailed calculations for tax processing. (This could be handled
directly between the programmers and the users.)

JAD is not a substitute for the management of people, and it should not
be wasted on simple tasks or detailed tasks best handled in smaller
meetings. To assure JAD is used effectively, you should set criteria for its
use. For example, the criteria might say:

JAD should be used for projects that:

• have high business priority or complex requirements
• require at least six MIS person-months
• involve users from more than one department
• are not involved solely with detailed programming specifications

---

To communicate these criteria to your company, you should produce a brief document that explains JAD and when to use it. Send this "manifesto" to key users (for example, system owners) and MIS managers. CNA Insurance, for example, distributes an eight-page glossy brochure to their applications development divisions. (Gill, 1987)

Properly used, JAD can save money. But it should not be used when less time-consuming management techniques would suffice. If you are not sure whether or not to use JAD on a project, ask yourself: How important is the project? How many users are affected? Would the project benefit from an impartial leader? How complex are the requirements? And, most important, is the project worth the concentrated effort of 12 participants for three to five days? Measure your answers against the criteria you have defined to decide if the JAD methodology is appropriate and cost effective for that project.

Now that we have covered the what, why, and when of JAD, let's look more closely at how the methodology works. The next part takes you through each step of the five JAD phases.

PART
2

THE JAD PHASES

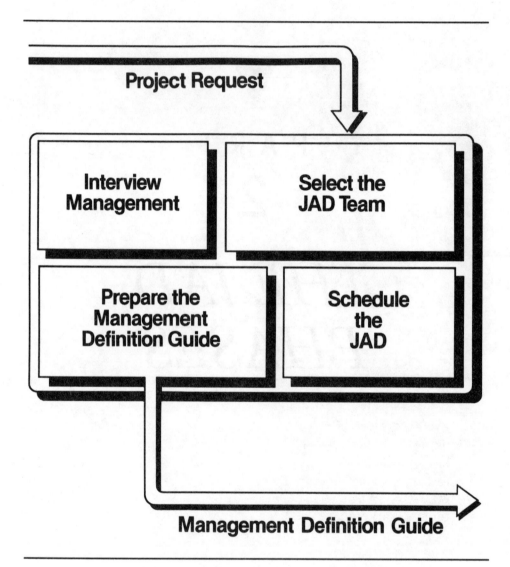

THREE ▷ ▶ ▶ ▶

PHASE 1: PROJECT DEFINITION

Where do projects come from?

- Some projects come from user requests. (The request form says, "We need this new enhancement to speed up processing and increase departmental operating efficiency.")
- Other projects are the result of extensive and convincing market research. (The concluding paragraph of a consultant's rambling report says, "For your company to remain on the leading edge in today's marketplace, you must update your old system to keep pace with the new generation of emerging technology.")
- And still other projects originate from a senior management whim. (The VP says, "Well, I was talking with my associates in Boston. They are converting all of their data processing systems to online real time. We should do the same. I think this new initiative, which I've outlined here on the back of this placemat, is a dandy idea. See if you can do something with it.")

No matter what the source of the project, whether it be a whim or a fully researched, unequivocally proven, do-or-die initiative, the project ends up in your hands. Your mission is to turn the project request (in whatever form it comes) into complete user requirements that can be handed over to the programming department.

To do this, you take the project through the following five JAD phases:

- Project Definition (Phase 1)
- Research (Phase 2)
- Preparation (Phase 3)
- The JAD Session (Phase 4)
- The Final Document (Phase 5)

This chapter covers Phase 1: Project Definition. First, it discusses how open issues and assumptions are handled throughout the project. Then, it describes in detail how to interview management, select the JAD team, produce the Management Definition Guide, and schedule the JAD.

OPEN ISSUES AND ASSUMPTIONS

Before we get into the actual steps of JAD, it is worthwhile to talk about two entities that are inherent to the methodology. Open issues and assumptions weave through all phases of JAD. They begin propagating as soon as the project begins. In early management interviews, for example, one user raises a concern that the leader notes as an open issue. At the same time, another user introduces or "assumes" a basic business decision about the system that the leader documents as an assumption.

Open issues are unresolved questions about the system design that can arise at any time during the JAD project. Following are examples of open issues:

- Will data entry be centralized in the Finance department, or will data be keyed in separately from user departments, as it is today?
- Will the system handle processing for our subsidiaries?
- Will we store fixed assets data in the existing inventory data base or will we set up a new area?

Assumptions are basic business decisions about the system that the participants have agreed upon and must keep in mind during the design process. Following are examples of assumptions:

- The first phase of the new system design will address only those functions required to satisfy government regulations.
- The system will handle fixed assets processing for the two acquisitions planned next year.

- The system will not be available to the field force until the end of next year.

As the Definition phase progresses, some of the open issues are resolved while others remain open. And new assumptions arise. Whatever their status, all issues and assumptions are brought into the JAD session for review.

During the session, discussions resolve some issues and identify new ones. At the end of the session, the team reviews all remaining open issues. All those that can be resolved are documented as assumptions. Even with all participants present, however, some issues may require additional research or may be slightly outside the scope of the project so as not to justify spending session time on them. Such issues are assigned to the appropriate people to resolve by a specific date. A coordinator is designated for issues assigned to more than one person. When the session is complete, all agreed-upon assumptions and unresolved open issues become part of the final document.

INTERVIEWING MANAGEMENT

Now, let the JAD begin. Before you compile, analyze, and document detailed system requirements, you need to identify what management wants from the application being designed. These higher-level requirements can be gathered in management interviews. The information will be used to:

- produce the Management Definition Guide. Written from the user perspective, this document includes the purpose, scope, and objectives of the project. (It is discussed later in this chapter.)
- select the JAD team.
- schedule the session.

Since you, the JAD leader, are interested in a broad scope of the system, you should interview managers who are directly responsible for the project. For example, you might interview the vice president responsible for the departments that will use the new system, the user department directors who will manage the daily use of the system, and the MIS project manager responsible for the system implementation.

Capturing the Information

There are nine special forms that can be used to capture the information required for the Management Definition Guide. These forms are de-

This form . . .	Answers these questions . . .
Purpose of the System	Why is the system being designed?
Scope of the System	Who will use the system and how often? What is the expected growth of the business? What areas of the company are affected by the system by either providing source documents or receiving reports? And what are the system interfaces?
Management Objectives	What do you want to gain from the system? What measurable results do you expect?
Functions	What functions will the system provide? What is the priority of these functions? For each one, is it "a definite must" or just "nice to have"?
Constraints	What limitations should be considered when developing the system? What are the deadlines? What is your budget? Are there any space limitations, security requirements, or government regulations? Can the data processing resources support an online system?
Additional User Resource Requirements	What are the user requirements for additional staff, equipment, and physical space?
Assumption (Pre-Session Version)	What decisions have already been made about the system?
Open Issue (Pre-Session Version)	What unresolved questions should be addressed prior to the session?
Participation List	Who should attend the session?

Figure 3-1 Forms for management interviews

scribed in the "Scribe Forms" section of Chapter 10. Meanwhile, Figure 3–1 shows a list of the forms and the types of questions they should answer. (This list will be more meaningful if you also take a look at the form samples in Appendix A.)

The best approach for extracting information from the users is to begin with a brief meeting with the user managers. Outline what type of information is needed in this phase and identify who will gather it. For example, one user may be able to define the functions the new system must provide while another has a better grasp of the overall objectives.

Give the forms to everyone in the meeting. Explain exactly what information goes on each one to reinforce what you are looking for. Then give the people a day or two to gather the information and fill out the forms.

Although the forms are a starting point for capturing the information, you need to go further. You need to reconvene to review the forms that the users filled in and probe deeper into certain areas. For example, a user may have written on the Management Objectives form, "The system will provide security from unauthorized access." But you need to know at what level. At the application level? Or will users have access only to specified screens within the application? You must have the users specify this information. The resulting clarified objective might be: "The system will provide security and access at the screen level."

To get the information in the format needed, you must guide, unravel, extract, and channel the discussion to the question at hand. You also need to restate what the users have said by using such phrases as, "Do you mean to say . . . ?" or "Let's see if I understand you correctly . . ." Then the user can confirm the information or correct you. You can add this new information to the same forms that the attendees filled out.

Additional Interview Questions

In addition to capturing the information described so far, there are other things you need to know. The following lists some other questions to ask during the interviews. Select the ones that relate to your project. The questions are organized into two groups: one for user interviews and the other for MIS interviews.

User Interviews

For first-time JAD participants, summarize the JAD team roles. (These are detailed in the "The JAD Team and How to Select It" section of this chapter.) Explain that you are interested in selecting the best possible people to fill these roles. Ask each manager:

- Which users should participate full time in the session to adequately represent user needs?
- Which areas of the company are indirectly affected by the system? Should someone participate in the session from these departments, or at least be readily accessible by phone?

To determine when the session will be held, ask:

- We have tentatively scheduled the session for May 7 to 11. Will these dates work for you?
- Full participation of the JAD team is critical to designing a quality system that works for you. Can you support us by committing the participants from your area to attend each day of the session?

Before concluding, ask:

- Are there any political issues that we should be aware of? For example, are there any people who might be difficult to work with because of their resistance to change? Are there any particular problems the users are having with the MIS department? (Saving these kinds of question until the end of the meeting, after you have developed rapport, will make the people more likely to discuss those sensitive issues you should know about.)

- Do you have any other comments or suggestions that might help us in preparing for the JAD session?

MIS Interviews

If you feel it would help to meet with MIS to complete the Management Definition Guide, set up an interview with the project manager. You might ask:

- Which MIS representatives should participate in the JAD session to provide technical information about the system?

For new systems design, ask:

- Will existing hardware be able to support the new system?
- Are you familiar with any software packages that handle this kind of application?

For enhancements to existing systems, ask:

- What are the technical problems with the existing system?
- Will adding functions to the existing system require reorganizing the data base?
- Do we have sufficient disk space and other hardware resources to support new enhancements?

To help determine when to hold the session, ask the same scheduling questions from the user interviews previously described.

With these interviews complete, you are prepared to select the JAD team, produce the Management Definition Guide, and schedule the session.

THE JAD TEAM AND HOW TO SELECT IT

The success of a JAD session depends largely on who is in the room for those three to five days of decision making. The following describes the roles of the participants and who should fill them. These descriptions include what each role does before, during, and after the session.

Executive Sponsor

This is the person in the user area with senior-level authority to make decisions for all aspects of the project. For example, we ran a JAD to define user requirements for a new Fixed Assets system. This involved participants from Accounting, Finance, Budget and Cost, Tax Planning, Facilities Management, and Administrative Services. The Vice President and Controller was an excellent executive sponsor because most of the departments represented in the JAD reported to her.

The prime responsibility of the executive sponsor is to make the words "management commitment" a reality. The importance of high-level commitment is not new. (Isn't that the song of every new project?) Management commitment for a JAD should begin with the executive sponsor. After all, this is the person responsible for all (or nearly all) the areas where the final system will be used. He or she has a vested interest in the project's success.

What if the logical person to select for executive sponsor has a weak personality or is an on-the-job retiree? This kind of person will not help your project and will not give you the management commitment you need. In this case, select the person who has the *most* authority to make decisions for the project *and* has the right personality. It is better to have an executive sponsor who can make some decisions for some parts of the project than to have someone who—even though he or she has the formal authority—will not make decisions at all.

Before the Session

The executive sponsor takes part in discussions with the JAD leader to define the purpose, scope, and objectives of the project.

To make the company's commitment known, the executive sponsor kicks off a preliminary meeting held the week before the JAD session. He or she briefly summarizes the project, its objectives, and its significance to the company. The executive sponsor also discusses his or her role in resolving open issues.

During the Session

Whether or not the executive sponsor attends the session depends on his or her involvement in the project. If involved in the day-to-day workings of a user area, the executive sponsor should attend. Usually, this is not the case, and he or she need not be there. The executive sponsor should, however, be accessible by phone throughout the session. There may be cases where an impasse is reached because two departments cannot agree. This requires a discussion with someone who can make decisions for both departments—the executive sponsor. Although we always plan for this contingency, we have never actually had to make such a call during a session. Before the session, however, we often consult with the executive sponsor on issues that affect more than one department.

After the Session

The only remaining job for the executive sponsor is monitoring open issues. After the session, open issues are resolved by the people assigned to them. Copies of the resolutions are sent to the executive sponsor, who then follows up on issues that are not resolved by the date set in the session.

JAD Leader

The key word that describes the JAD leader is *impartial*. This person, who guides the team through the complete JAD process, should be objective, unbiased, and neutral. He or she should have no vested interest in how the final product is designed except that it run effectively, efficiently, and meet user needs. Ideally, the leader comes from neither the user area nor the programming department. For example:

- Our company has a group separate from programming (but still within MIS) called Business Systems Engineering. This group is responsible for running all JAD projects.

- Other companies have introduced IBM's concept called the *Development Center*. Here again, a separate staff is hired whose sole purpose is to implement tools and techniques such as JAD that increase productivity in the development group.

Both these reporting structures support impartiality. In smaller companies (and even in some larger ones), a liaison group between programming and user departments may not exist. Then the leader more likely comes from programming. When acting as leader, however, he or she is there to lead the session rather than to represent MIS views.

In any case, the leader is neither designer, analyst, nor project manager. He or she is a guide, a facilitator. During the session, the leader moderates and steers the participants through what can be a rugged terrain of decision making. He or she should have no stake in whether a data element is 10 characters or 20, or whether the report is sorted by customer or location. Those decisions are up to the users. The leader simply provides a forum in which the decisions can be made. In other words, JAD leaders don't write the songs; they simply run the recording studio.

What Makes a Good JAD Leader?

Unlike most other JAD roles that are assigned only for the duration of the project, the leader can be hired into a full-time position whose main responsibility is doing JADs and planning for them. When JADs are conducted less frequently, the leader can come temporarily from a traditional systems analyst or project manager position.

Some companies use outside consultants to lead JADs. Consultants usually bring everything needed for the session, but all for a rather significant fee. This is cost-effective only if you plan to do no more than one or two sessions per year.

When selecting a leader, you need someone with an energetic, outgoing personality who can:

- organize on a project level
- communicate well
- lead groups
- summarize discussions
- steer groups away from tangents and unnecessary details
- be sensitive to group dynamics and company politics

The leader should have a good understanding of data processing concepts. Finding someone who is familiar with both structured analysis and data base design is a definite plus. Knowing the business application being designed gives depth to the leader's effectiveness (relevant discussion can be more easily distinguished from extraneous digression). However, it is not essential that the leader know the application in detail. In fact, too much knowledge can cause a leader to become dictatorial in a role that should remain neutral.

In a recent JAD user's group meeting in the Hartford, Connecticut, area, the attendees were asked "What do you like about being a JAD leader?" The responses were interesting. Some said they were attracted to the power and the control. They had always enjoyed the process of turning a floundering meeting into a productive session. Others were more interested in being "part of the solution." Leading JADs allowed them to take on the role of "professional middleman" between users and MIS. Others enjoyed learning about different parts of the company through the variety of JAD topics. And still others were hams at heart and felt JAD leading offered a corporate stage for their talents.

Before the Session

The leader conducts interviews with the executive sponsor and with management from user departments and MIS. This is when the project purpose, scope, and objectives are identified to build the Management Definition Guide (described later in this chapter). Then, working with users and MIS, the leader gathers information on work flow and system requirements. This is compiled into a Working Document to use in the session (more about this document in Chapter 5). The leader also prepares a script for the session and visual aids such as flip charts, magnetics, and overheads.

During the Session

The leader guides participants through the agenda, remaining impartial and aware. He or she flags tangents and keeps the session on track. The leader is sensitive to what interests are being represented to assure that no one area is dominating. The leader determines when agreements are reached and prompts the scribe to document the agreements. When open issues arise, the leader writes them on a flip chart to review at the end of the session. And just before the participants reach their overload point, the leader calls a welcomed break.

After the Session

Once the session is over and agreements have been documented, the leader oversees the creation, review, and distribution of the final document. When appropriate, he or she leads the post-session review meeting.

JAD Support Person

This optional role helps when you run JADs for large projects with tight time constraints. This person handles many tasks normally done by the leader, such as preparing the visual aids, the Working Document, and the final document. During the session, he or she sits next to the scribe, monitoring what is being recorded, to aid in later translating the scribe's notes into the final document. Having this person follow the project from the start is beneficial so that, in the unplanned absence of the leader, the support person can take over leading the session. In fact, during one of our major JAD projects, the leader fell off his bike on the way to work and injured his ribs. The support person had to fill in . . . the JAD session must go on!

Scribe

In the session, one person is designated to record all decisions. In a twentieth-century revival of a somewhat archaic word, this person is called the *scribe*. This role is critical to a successful JAD. Just as in the days of robed monks, quill pens, and illuminated manuscripts, scribing is an art. Unfortunately, the modern-day role is not so romantic. You might hear comments like, "I'm a programmer, not a scribe!", because people sometimes see it as grunt work, secretarial work, and definitely *not their kind of* work. But, on the contrary, scribing is a key role that can demand serious analytical skills.

The notes taken by the scribe evolve into the final document. This process is simplified by a series of forms used to record the specifications. For example, when a new data element is defined, the scribe documents it on the Data Element Description form, which includes space for data element name, length, format (alphabetic, numeric, or alphanumeric), and a complete definition.

What Makes a Good Scribe?

The scribe comes from either the MIS or user area at staff level. You need someone who:

- has a good working knowledge of the system. That is, he or she understands it enough to be able to capture decisions accurately.
- has good analytical skills. (Programmer/analysts and supervisors from user departments often make excellent scribes.)
- can take notes well and extract from discussions. Excellent skills, however, are not required here since the standard scribe forms, and prompts from the leader, eliminate the need for synthesizing or interpreting.
- has clear handwriting (doctors need not apply).

Selecting someone who will be using or programming the final system yields a good educational by-product. Scribing is an excellent way for someone (who will need to be trained anyway) to learn about the system before it is programmed.

Before the Session

The leader meets separately with the scribe to talk about the scribe's role in the session and to discuss what forms should be used to capture the requirements.

During the Session

As the session progresses and decisions are made, the scribe does not need to determine when or what to write. The leader prompts when to document and the scribe forms show what to document. Chapter 6 details this process and Appendix B shows actual scribe form samples.

After the Session

The scribe reviews notes with the leader. Depending on available re-sources, the scribe can help prepare the final document.

Full-Time Participants

Full-time participants include all the users and MIS people involved in making decisions about the system design. The users know what they need. The MIS people know how these needs affect the computer environment.

> JAD encourages corporate consensus with each participant bringing his or her own experience and vision to the table. It is, further, an opportunity to get business requirements and system specifications right the first time. In so doing, the bank's information needs are served better and more quickly.
>
> —J. C. Grant
> Executive VP, Operations and Systems
> Bank of Canada (Brown, 1987)

Commitment is essential. Full-time participants are just that, *full-time*. Everyone must attend every day of the session. If a person returns having missed even one session day, you have the following problems:

- Half the group spends valuable time updating that person on what was missed while the other half becomes bogged down, bored, and frustrated because they "went through all that yesterday."

- Decisions will probably be revised (taking even more time) because that person's needs were not known, let alone considered.

JAD is a cumulative process, like learning algebra. You have to be there today to understand what happens tomorrow. And you cannot be the "lone ranger." You are dependent on what others contribute, and they are dependent on you. One person's comment leads to another person's suggestion which leads to a final decision that works for everyone involved. Therefore, full-time participants must attend the entire session.

Who Are These JAD Participants?

For success in the session itself, and ultimately with the system design, participants should be knowledgeable about the application and willing to discuss their opinions, not argue them. A person's rank in the company should not give that person an advantage in the session. *All participants are equal.*

Users range from lead clerks to supervisors to upper management. Clerks and supervisors are concerned with how the new system will support their day-to-day work. (Are the screens easy to read? What are the default values? And how many keystrokes will it take to get from one screen to the next?) Management is concerned not only with ease-of-use but also with what the system produces. (Do the reports give us the information we need? Are statistics provided to make management decisions?) At least some users at the session must know the system completely and have the authority to make decisions. The words "I'll have to check with my boss," should not be heard in a JAD session.

MIS participants range from programmers to project managers responsible for the system being designed. They provide information on existing systems and technology. They can determine the feasibility of design requests, estimate costs, and suggest other approaches when necessary. Also, their presence ensures that they understand user needs and can translate the resulting decisions into an effective system design.

When multiple company locations (for example, subsidiaries or divisions) are directly affected by the new or enhanced system, those locations should also be adequately represented.

Determining who should attend sessions requires a balancing act. You want enough people to have full representation and decision power in all the user areas that are directly affected. At the same time, you want to keep the session small enough to be productive. As every business person knows, beyond a certain number, the productivity of meetings is inversely proportional to the number of attendees. So try to find the balance between too many people and not enough. Eight to fifteen people is about

Title	JAD Role
Vice President, Controller	Executive sponsor
Senior Business Systems Analyst	Leader
Director, Finance	Team (User)
Supervisor, Finance	Team (User)
Data Entry Clerk, Finance	Team (User)
Director, Budget and Cost	Team (User)
Director, Tax Planning	Team (User)
Director, Purchasing	Team (User)
Director, Facilities Management	Team (User)
Project Manager, Financial Systems	Team (MIS)
Senior Programmer/Analyst, Financial Systems	Team (MIS)
Programmer, Financial Systems	Scribe (MIS)

Figure 3-2 List of JAD session participants

right in most instances. More are required for large systems that affect multiple departments.

A good ratio is five users to two MIS people. Figure 3–2 shows a sample participant list for a typical session.

Users on Call

These people are affected by the system design, but only in one particular area. In other words, most of the session is not relevant to these users. They should participate only when and if their expertise is required. For example, someone in the law department might attend two hours on Wednesday when legal implications are covered. Or they might be consulted by phone. In either case, they need to be accessible on the session days when their expertise is needed, as determined by the agenda.

Observers

Sometimes people ask if they can sit in on a session. Perhaps they are training to become a leader, or just want to know what this "JAD mania" is all about. Observers can receive memos and other information sent out before the session, but they should not actually participate in the session. Muzzle them. Speak with observers beforehand to make sure they understand their role (or lack thereof).

THE MANAGEMENT DEFINITION GUIDE

Management Definition Guide. Could a more mundane title for a document ever be found? Doesn't it sound like something to read when you can't sleep and have already tried counting sheep?

Despite the dryness of its title, the Management Definition Guide contains important substance. It defines what management wants from the application being designed. It is written from the user perspective and includes the purpose, scope, and objectives of the project. It communicates management's direction and commitment.

Contents of the Document

The content of this 10- to 20-page document comes from interviews with user and MIS management. How these interviews take place was covered earlier in this chapter. The following describes what you gather in these interviews that comes together to build the Management Definition Guide.

The following example is based on the project described in Chapter 1, where your company wants to design a robot to cater meals for company functions. This is a good example to show what each section of the document accomplishes and to clarify such concepts as purpose, objectives, and functions. Once these concepts are clear, you then need to know how to document each section of the Management Definition Guide for the actual projects you are faced with. Chapter 11 shows a complete sample Management Definition Guide for a typical business application.

Title Page

The title page begins with the document's prosaic identifier, "Management Definition Guide," followed by:

- the project name
- the publication date
- the names of contributors
- the company name

MANAGEMENT DEFINITION GUIDE
ROVING ROBOT PROJECT

April 1, 1990

Karen Alexander
Sara Balderston
Joseph Darcy
Jean Mugler
Beverly Penn
Robert Zimmerman

Permanent Assurance Company

Preface

This summarizes how the Management Definition Guide fits into the JAD methodology. For example:

PREFACE

This document results from the first phase (Project Definition) of the JAD methodology. Information gathered in interviews with the executive sponsor, key user, and MIS managers has been compiled into this document, the Management Definition Guide.

Table of Contents

This lists the document sections and their page numbers. You can divide more extensive sections into subsections, as shown with the items listed under the heading *Functions* in the following sample.

TABLE OF CONTENTS

Purpose

This tells *why* the system is being designed.

PURPOSE

The company faces increased demands for catering services at company functions. Senior management has determined that a Roving Robot system can fulfill these culinary needs in a cost-effective manner. The robot will prepare and serve meals for breakfast and luncheon meetings as well as cater larger social functions, usually held in the evenings.

Scope

This tells *who* will use the system and how often—which departments, divisions, offices, and the geographic locations of the users. It lists the areas indirectly affected by the system, those that will either provide source documents or receive reports. It quantifies the expected growth of business in the area using the application. And it describes what other existing computer systems are affected.

SCOPE

The Roving Robot system will be administered by the Corporate Services department to provide food services throughout the company. Ninety percent of the robot's services will be used by the following departments:

- Communications
- Sales and marketing

The expected volume of requests for catering can be estimated based on last year's volume, which totaled:

- 55 breakfast meetings averaging 15 people per meeting
- 80 luncheon meetings averaging 20 people per meeting
- 30 evening functions averaging 40 people per function (20 events were for drinks and appetizers; 10 included full-course dinners)

We expect these requests to increase 10% this year, 10% next year, and (due to the planned subsidiary acquisition) 20% the year after that.

The robot system will receive documents from the Information Desk (for catering requests) and will send documents to the Purchasing department (to handle major food purchases).

It will automatically interface with the Chargeback system to charge cost centers for its services.

Management Objectives

This tells *what* management expects to gain from the system. Try to quantify the objectives. Many projects profess to "increase productivity," but what does this really mean? Include such information as expected increases in sales volumes, decreases in costs, changes in inventory levels, and measurable improvements in customer satisfaction. Quantified objectives allow you to better determine if the resulting system meets those objectives. In the following example, notice that two out of the four objectives are quantified.

MANAGEMENT OBJECTIVES

Management objectives for the Roving Robot system are to:

- reduce by 20% the cost of having meetings catered by outside food services.
- increase reliability. The company has had a long history of bad luck with local caterers. With our own robot, we will have more control over the kind of service provided.
- decrease by 50% the Corporate Services work load of having to process meal requests, contact caterers, and call out for pizza when food does not arrive.
- increase the quality of food served at company functions. Satisfied, well-fed clients translate into more business for the company.

Functions

There is a subtle distinction between functions and objectives. Functions tell what the system will do (prepare meals) while objectives tell what management will gain from these functions (catering costs are reduced).

FUNCTIONS

The Roving Robot system will do the following:

Plan meals. The robot prepares meal plans based on:

- which meal (breakfast, lunch, dinner, or snack)
- type of attendees (staff, management, or clients)
- season (summer, fall, winter, or spring)
- desired ambiance (formal, informal, or festive)

After processing this information, the robot prints the following reports:

- menu
- shopping list, including what to buy and how much

Prepare meals. After the meal plan has been verified and supplies purchased, the robot assembles the ingredients into the appropriate cuisine. The robot will:

- identify equipment such as pots, pans, woks, and egg beaters.
- calibrate volume ranges from one pinch to six gallons. (It also handles metric conversions.)
- distinguish between food entities such as lettuce heads and artichoke hearts.

Serve meals. The robot transports hot and cold meals to the meeting location.

Clean up afterwards. After dessert, the robot clears the table, washes the dishes, puts everything away, and prepares doggie bags on request.

Constraints

This describes limitations to consider when designing the system. The most common constraint is time (the application must be installed by April 1). Others include money (we cannot exceed a $500,000 budget), space (we have only 1200 square feet to hold the equipment), system limitations (no additional hardware will be purchased), and government regulations. Sometimes time zone considerations are handled in the constraints section.

CONSTRAINTS

Constraints are:

- The robot must be ready for use by December 10 of next year to serve the annual Christmas party.
- The robot must perform all food preparation in the 800 square foot space now occupied by the copy center.
- The cost for the complete robot system cannot exceed $500,000.

Additional User Resource Requirements

This identifies additional user requirements such as people, physical space, hardware, communication needs, and data security.

ADDITIONAL USER RESOURCE REQUIREMENTS

To implement the Roving Robot system, the following additional resources are needed:

- one part-time taster to shop for the food specified on the Shopping List report.
- one full-time worker to maintain quality control and service the robot.
- a closet with a lock to store the robot.
- a full complement of cooking equipment and supplies. (This list, which should include everything from a carrot-crunching cuisinart to a stainless steel garlic press, will be finalized in the JAD session.)

Assumptions

Any general business decisions that emerge from the management interviews are included here. The list of assumptions will grow before and
during the JAD session.

ASSUMPTIONS

Assumptions about the Roving Robot system are:

- The Corporate Services department will manage all services relating to the robot.
- Robot services will be charged back to the departments using them.
- Meals planned by the robot must meet corporate health standards (no junk food allowed).

Open Issues

This includes questions that have come up before the session that need to
be resolved.

OPEN ISSUES

Open issues are:

- How will we handle the conflicting food and beverage preferences among departments? For example, there are several opposing views on what should be included on the wine list.
- How will we accommodate special requests, such as for vegetarian and kosher foods?
- What will we do for backup when the robot is out of service? Should we resort to caterers and pizza parlors as we do today, or try to prepare meals ourselves in the newly constructed robot kitchen?

JAD Session Participants

This lists all the participants who will either be attending the session or be
on call during the session. It also includes the executive sponsor. As
interviews progress and the participant list comes together, you may run

into situations where you feel the wrong person has been selected to attend the session. There are a couple of ways to deal with this. First, make sure the manager understands the impact this person will have on the design effort. Do this in a preventive way, before the participants are officially selected, to assure the manager gives serious thought to the selection. If, however, that wrong person still makes it onto the participation list, get an alternative recommendation from the executive sponsor. Then try to persuade the original manager of this better choice.

When the participants have been identified, list them along with the departments they represent, their mail code or address (for distribution), and their role in the session. This preliminary roster will probably change somewhat before the actual session.

JAD Session Participants

Participants for the session include:

Name	Department	Mail Code	Role
Karen Alexander	Sales & Marketing	4	Team
Sara Balderston	Communications	3	Team
Joseph Darcy	Corporate Services	7	Executive Sponsor
Peter Lake	Facilities Management	2	Observer
Jean Mugler	MIS	9	Team
Beverly Penn	Communications	3	Team
Renee Rudolph	MIS	9	Leader
Terry Silver	MIS	9	Scribe
Robert Zimmerman	Corporate Services	7	Team
Kenny Zonies	Sales & Marketing	4	Team

Send the completed Management Definition Guide to the contributors for their review. Give them a specific amount of time to return their comments. Two or three days should suffice. Then, send the revised document to all participants so they can read it before the session.

SCHEDULING THE JAD

The time required to complete a JAD depends on the scope of the project and, of course, the time constraints. Some projects can take several months to define rather straightforward specifications. Others must be done in less time. One project in our company involved adding major

enhancements to five separate application systems. Using JAD, we defined the user requirements in four months. This involved user and MIS
interviews, 25 back-to-back full-day sessions, and six final documents
which defined data elements, screens, reports, calculations, and data base
requirements.

How Much Time Does Each Phase Take?

Figure 3–3 shows how much time to allow for each JAD phase. Some time
estimates are shown in ranges. This indicates that the step varies depending on the project scope and level of detail required. For example, in
phase two, researching data elements, screens, and reports has a range of
two to four days. If you are documenting a relatively small system (with
only 15 data elements, 5 screens, and 3 reports), then two days will
probably suffice. But if you are documenting the requirements for a
larger system (with 50 data elements, 20 screens, and 15 reports), then you
will need more time.

As you can see from Figure 3–3 most of our JAD projects last between
five and ten weeks.

Other JAD users and consultants often give lower time estimates be-

No.	Phase	Step	Work Days
1	Project Definition	Interview management Produce the Management Definition Guide Schedule the JAD	1 to 3 1 to 3 1
2	Research	Get familiar with the existing system Document work flow Research data elements, screens, and reports Prepare the session agenda	1 to 4 1 to 5 2 to 4 1
3	Preparation	Prepare the Working Document Prepare the JAD session script Prepare overheads, flip charts, and magnetics	2 3 to 5 1
4	The JAD Session	Hold the session	3 to 5
5	The Final Document	Produce the final document Participants review the document Hold the review session Update and distribute the final document	3 to 10 2 1 2

Figure 3-3 Time required for each JAD phase

cause they do not include the up-front work to the extent we do. They may, for example, omit documenting work flow or may not even prepare a Working Document. In these cases, a JAD project involves much less time but results in less comprehensive specifications.

Half-Day or Full-Day Sessions?

We began holding workshops in full-day sessions, but soon realized the advantage of half days. No matter what the setting, full days are strenuous for everyone, particularly the leader. By mid-afternoon, unless you can add scintillating humor and spellbinding enchantment to the subjects of data elements and screen design, you will probably lose the participants to cognitive burnout. Productivity diminishes and afternoon caffeinated cola can recharge only the most resilient.

With half-day sessions, participants are more attentive and less preoc-cupied with what might be happening (or not happening) back at the office. They still have the non-JAD part of the day for other respon-sibilities. Also, when people need to get information about a particular item discussed in the session, they have time to make contacts and do research in their own environment rather than making hurried phone calls during a break or staying after normal work hours.

Leaders benefit from half-day sessions as well. They have time to review scribe forms and research questions. They can prepare flip charts, magnetics, or overheads showing, for example, the new screen designs that may have emerged in that day's session.

Another advantage of half-day sessions is scheduling. Bringing together a group of key company people for up to five days straight is challenging enough, let alone trying to arrange all *full* days. With half days, the participants still have some time for their other work.

Sessions are more productive in the morning than in the afternoon. Early in the day, people are more awake and ready to get started. Sessions can be held from 8:30 to 12:30 with a 15-minute mid-morning break.

IBM recommends full-day sessions. This approach does have its advan-tages: momentum is more easily maintained and the session is completed sooner. Nevertheless, we have found that half days work much better.

Being Off-Site and Inaccessible

Sessions can be held anywhere you have sufficient space, magnetic boards, and a place to project overhead images. However, holding them off-site in a separate building has several advantages. Of course, you pay the charges for the room, coffee, and snacks. And participants must take time to travel to the off-site location. But these inconveniences are worth the benefits gained.

The main advantage of being off-site is simply stated: *out of sight, out of mind.* Being away from work prevents distractions such as co-workers who just want to "check on something" or pass along messages that you might want to follow up on.

Another advantage is partly psychological. Most people enjoy getting away. The change in environment frees the mind to focus more completely on the project at hand. This makes for a more productive session.

When evaluating meeting rooms, whether on-site or off, look for:

- *Large tables.* Participants like to have many items in front of them, such as JAD documents, program listings, notes, cups of coffee, a serving of Danish, and a name card.
- *Audio-visual accommodations.* If you want to rent equipment, check the models and prices. Evaluate the room for projection conditions. Is there enough space to project a sufficiently large image?
- *Board space.* The more board space the better. To use the magnetic techniques described in the "Visual Aids" section of Chapter 5, you will need a white magnetic board. This allows you to write with colored pens as well as to move magnetic shapes around the board when designing screens and reports.
- *Storage space.* Having a locked area allows you to store your supplies and equipment overnight.
- *Food services.* Are refreshments available? For example, it is nice to have coffee and pastry for morning sessions and soda for afternoon sessions.
- *Flexible scheduling.* Can sessions be scheduled on consecutive days (to keep the momentum going) and in the same room (to save setup time)?
- *Comfort.* Good ventilation and climate control are also important.
- *Convenient location.* Being close to the office is nice, but not as important as the other considerations.

Finding a Room for the JAD Session

In our years of running sessions, we have not used a regular hotel facility yet. They are relatively expensive and generally do not have good board space. Instead, try buildings that lease office space to small businesses. They often have available meeting rooms. Colleges and universities are sometimes suitable, but the challenge is finding a room that has tables (most have only desks) and that is available for several consecutive days.

Also, some colleges rent only to nonprofit organizations. If you do find a room on a campus, you will probably have excellent board and wall space, not to mention the creative environment and nostalgic effects of a college campus. (Remember when success was maintaining a 3.5 cumulative average and pressure was starting a 25-page term paper the night before it was due?)

If you cannot find a meeting room on a campus or in a place that leases office space, you can rent from a hotel. For rooms with little or no board space, you can rent or buy portable magnetic boards from audio-visual equipment or office supply companies.

For more suburban locations, you might contact a neighboring company about available rooms. Or you could inquire into local public or parochial schools.

When to Hold the Session

Determining when to hold the session depends on whether or not the target date for implementing the system has been set. Target dates are

---

August 20, 1990

To:	JAD Participants
From:	Renee Rudolph
Subject:	The JAD Session

The Roving Robot system JAD session will be held in the Liberty Building at 3600 Market Street. To determine the complete specifications for the system, it is vital that all participants attend.

Location:	Liberty Building (Room 101) 3600 Market Street
Dates:	September 26 to 30 (8:30–12:30)
Phone:	Messages can be left at 555-2200.
Public Transportation:	Market-Frankford Elevated A-train to 34th Street
Parking:	Across the street.
Snacks:	Coffee, tea, and danish.
Copies:	(executive sponsor) (your boss) (others who should be advised)

Figure 3-4 JAD session memo

either flexible (the users have not set a firm date) or mandated (upper management or "Uncle Sam" says install the system by July 1). The following describes how to handle each type of target date.

For *flexible* target dates, management waits until after the session to determine a realistic implemention date. This is by far the best way to set target dates. Once the session is complete, the complexity of user requirements is known, which enables management to set more realistic target dates. Also, with flexible dates, you have time before the session for more detailed analysis of work flow, screens, reports, and processing requirements.

For *mandated* target dates, plan the project working backward from the final date. For example, you might have a mandated target date of January 1. Working backward from that date (factoring in the time required for coding and testing), you can plan when the JAD session must be held.

Once you set the session dates and select the meeting room, send a memo (from you, the JAD leader) to the session participants with copies to:

* the executive sponsor
* your boss
* any others who should be advised

Figure 3–4 shows a sample of this memo.

Having completed the Project Definition phase, you are ready for Phase 2: Research.

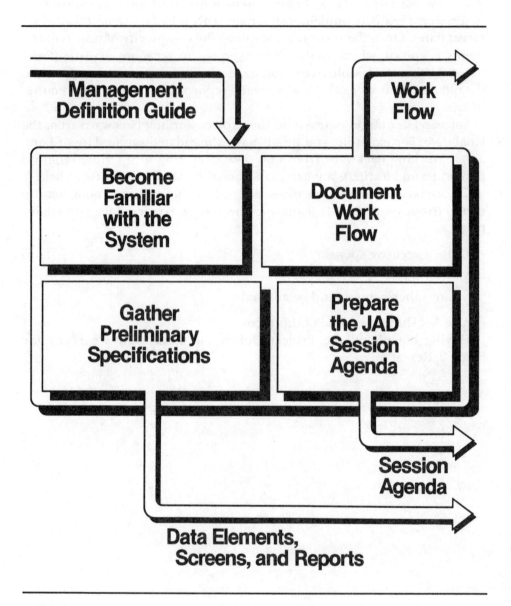

FOUR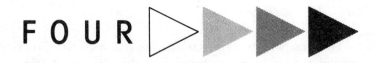

PHASE 2: RESEARCH

Having selected the JAD team, prepared the Management Definition Guide, and scheduled the session, you are ready to:

- become familiar with the system
- research and document work flow
- gather preliminary specifications for data elements, screens, and report
- prepare the session agenda

The information gathered in this phase goes into the Working Document prepared in Phase 3.

GETTING FAMILIAR WITH THE SYSTEM

From preparing the Management Definition Guide, you have a good idea of what management wants from the system. But you may not have even an inkling of how the current system actually works. In fact, there may be no current system at all. For these situations, familiarize yourself with whatever specifications have been prepared so far for the new system. In most cases, however, some kind of system is in place, whether automated or manual.

The best way to familiarize yourself with the functions of the current system is to set up meetings with the users and MIS. You need to learn enough about the business to be comfortable with the terminology and buzzwords. And you need to know enough about how the current system operates so that there are no "surprises" in the session.

Meeting with the Users

Hold the meeting in the users' area. Meet with those who actually work with the system on a daily basis. The ideal person is a working supervisor who has performed all of the system's functions.

1. *Observe the working environment.* Take a user-guided tour through the workplace. Some leaders take photographs of the key user areas. Then during the session, they project the slides to allow participants to see a visual image of the area being discussed.

2. *Observe the work flow.* How do they do their jobs? If there is an existing system, log on to it. Spend some time with the main menu and submenus, as these often summarize the system functions. Ask the users to perform these functions as they normally would. Where do they get the information they need to accomplish their tasks? What is the volume of activity? Have them show you the screen flow and how the screens branch. Ask them to identify which screens are for updating information and which are for display only. Have them enter some real data. Discuss the processing; in other words, what happens to that data?

3. *Review the output.* Since the output is usually in the form of printed reports, tell the users well before the meeting that you are interested in seeing those reports. Find out the format and frequency of reports.

4. *Discuss system changes.* Do they like using the current system? Are the screens easy to read? What are the problems? What causes time delays? How could the system be changed to make their jobs easier? What new functions need to be added to support the new requirements?

Meeting with MIS

Your goal for this meeting is to get a more technical overview of how the system works. You want to fill in the gaps from a data processing perspective. Therefore, arrange the MIS meeting not only with the project manager, but with the programmer analyst who works directly with the programs.

1. *Discuss the MIS view of the system.* Review system flow charts. If you need a more detailed technical view, you might want to become familiar with the program descriptions, files, and daily transactions. Ask to see the data base design. (For example, have them show you the Bachman diagrams.)

2. *Discuss the project in general.* Ask about the project history. Ask their opinions about the system. What changes do they think should be made? What are the current weaknesses in the system and how can they be addressed? Do they think the new requirements will be easy to implement?

Does the JAD Leader Have to Know Everything?

As you gather information you, the JAD leader, will come face to face with the question, *how much do you really have to know?*

The more of a perfectionist you are, the more you are burdened with the quest to know everything. There may be a compulsive part of you that wants to learn every detail about every function in the system. You want to know it as a user would.

The best thing you can do is to divest yourself of this self-imposed burden right now. The JAD leader does *not* have to know everything. By realizing this, you will save yourself a lot of time and energy plodding through reams of source code, pages of calculations, and other details that can only bog you down.

Keep your perspective on a higher level. Yes, you need to know the system functions. You need to know generally how they are used. But you do not need to know every value that a field can have. You are not preparing to fill in for fallen users. You are preparing to lead a session in a room full of participants who already have their heads filled with such details as how the program performs calculations and which codes are valid for a given field. These system experts will handle the details.

One application system could have a project team of ten people whose sole purpose in life (so it seems) is to know the definition for every data element, the calculations, and other nuances. You, the JAD leader, could be dealing with six major systems in one year. How could you possibly learn all the details? Your job is to keep sight of the big picture, run the JAD using the resident experts, and move on to the next project.

DOCUMENTING WORK FLOW

There are several ways to document work flow. Some companies use flow charts, some use data flow diagrams, and some just write it out in words such as, "This information goes here and that information goes there." We use data flow diagrams. This technique, which comes from the world of structured analysis and design, documents the flow of information be-tween business activities. The most common methods for creating data flow diagrams are those of Yourdon and Gane & Sarson. We use a combination of both.

To create these diagrams, we use a CASE (Computer Assisted Software Engineering) tool. Although such tools make the job easier, they are not absolutely essential. You can still do the job without them. But they do help when it comes to revising and analyzing the diagrams. As the CASE salesperson will tell you, people are reluctant to change system designs because it means having to redraw all those diagrams. CASE tools make this easier and can save a lot of pencil lead and eraser crumbs. Also, most of these tools have a data dictionary—a central repository for storing information about the diagrams for later use. The *"Can CASE Tools Help?"* section of Chapter 10 discusses how CASE tools can be used in JAD.

Parts of a Data Flow Diagram

To set the stage for describing how to identify work flow and create data flow diagrams, it will help to talk about the kind of information you are dealing with. Data flow diagrams contain four kinds of information:

- data flows
- processes
- data stores
- external entities

The following describes each of these parts. Figure 4–1 shows a data flow diagram that uses "Pay Bills" as an example. Refer to this diagram as each part is described.

- *Data flow.* This is the information that moves through the system. For example, your telephone bill is a data flow. It contains pieces of data moving from the telephone company to you. Data flows can be drawn at any level of detail. Just as your telephone bill is a data flow, so is each piece of information on it, such as your account number, name, address, and the amount due. In the diagrams, data flows are shown as arrows:

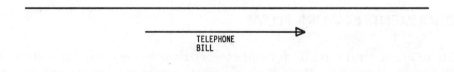

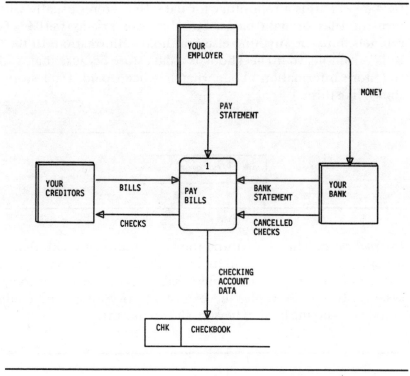

Figure 4-1 Data flow diagram

- *Process.* This is what causes data to change. Data moves into a process, is somehow changed, and then moves out of the process. For
  example, the telephone bill moves into a process called *Pay bills,*
  where the data is reorganized into the form of a personal check and
  sent back to the telephone company. The best way to make sure
  something is a process is to see if the data has, in fact, changed. If
  someone receives data, does nothing with it, and passes it along in
  the same form, then a process has not occurred. Processes are shown
  in a rounded rectangle with an accompanying number that uniquely
  identifies the process (more about these numbers later).

- *Data store.* This is a repository for data. Data stores usually take the form of files or data bases, but they can just as well be filing cabinets, bins, or anything else that holds information. In the "Pay Bills" example, your checkbook is a data store because that is where you store information about what has been paid. Data stores are shown like this:

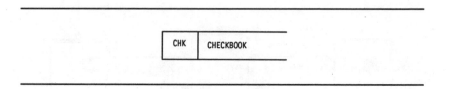

- *External entity.* This is anything the system interacts with that is not actually part of that system. In a typical business example, external entities might include customers, sales agents, or other computer systems. In our example, the external entities are your creditors, employer, and the bank. They are shown like this:

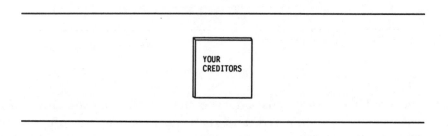

Figure 4–1 (the data flow diagram for paying the bills) includes a process, several data flows, a data store, and three external entities.

There is another aspect of data flow diagramming that you should know about at this time. Processes can be *exploded* into more detailed levels. For example, in Figure 4–1, the process "Pay Bills" can be exploded into several subprocesses such as sort the bills, write checks, and balance the checkbook.

To identify the level of explosion, the processes are identified with numbers. If the "Pay Bills" process is process number 1, it could explode into the following processes:

1.1 Sort Bills

1.2 Write Checks

1.3 Balance the Checkbook

For some other project, process number 2 might explode into a diagram containing these processes:

2.1

2.2

2.3

Then process 2.1 could, in turn, explode into these processes:

2.1.1

2.1.2

2.1.3

2.1.4, and so on.

Figure 4–2 shows how these exploding processes are named.

What we have shown is one way to diagram work flow. You may already be using this method or some other approach that works for you. This brief description of data flow diagrams is just the tip of the iceberg. Complete books have been written on this and other structured analysis techniques. If you would like to learn more about data flow diagramming, refer to some of the structured analysis and design books listed in the Bibliography. For now, we have given you an overview because it ties into the following discussion.

Capturing the Work Flow

Now, how do you get the work flow out of the users' heads and onto paper? To do this, you lead a series of small design meetings. You use overhead

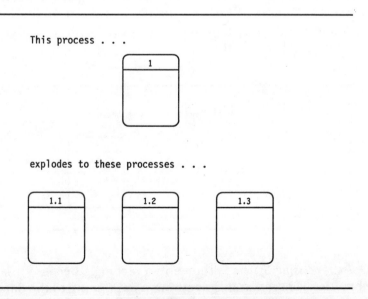

Figure 4-2 Naming processes in data flow diagrams

transparencies, a blackboard, flip charts, and those little yellow remov-
able self-stick squares of paper that people use to stick notes on other
sheets of paper.

The following description uses the data flow diagramming techniques
we have just summarized. This description is not meant to make you
fluent in data flow diagramming. Instead, it shows you how to *capture work
flow*, whether you use data flow diagrams, flow charts, or whatever.

This four-step method uses an example involving the design of an
Order Processing system. This example will continue throughout the
book as we describe the complete JAD methodology. For now, we are
concerned only with capturing work flow. Following is a description of
how it works.

1. Identify the First Level of Work Flow

Arrange a meeting with the two or three people who really know the
flow of work. They are usually a user manager or supervisor and perhaps
someone from MIS. Also, have a scribe attend to assist in documenting
decisions. You will see the importance of this in a moment.

Open the meeting by explaining how the work flow will be dia-
grammed. If you are using data flow diagrams, for example, describe the
technique as we did beginning on page 56.

Start with the first level of work flow. This is the *context* diagram, that is,
the highest-level diagram that shows the scope of the system and the net
flow of data in and out of it. In our example, the scope of the system is
"Order Processing." Therefore, the context diagram is named *Order Processing System Context*. Write that name at the top of the board. Draw a
rounded rectangle in the middle of the board with the system name and a
zero for its number, like this:

Order Processing System Context

```
        ┌─────────────┐
        │      0      │
        ├─────────────┤
        │ ORDER       │
        │ PROCESSING  │
        │ SYSTEM      │
        └─────────────┘
```

Order Processing System is shown in a rectangle because it is a process in
the broadest sense; that is, something is happening to the data.

Next, identify the data flows. Ask the group, "What comes into the system and what goes out? We are not concerned with details here. If we have one line of data flow coming into the process and one going out, that is sufficient." Keep the diagram as simple as possible.

The group will probably ignore your plea for summary information and try to give you the whole enchilada: "Well, we get requests for new orders, order changes, order cancellations, address changes, and . . ." After you ask them to summarize all that into just a few data flows, they suggest that *order information* comes into the system and the following goes out:

invoices

billing information

packing slips

shipping labels

reports

Add these flows to your diagram. Now it looks like this:

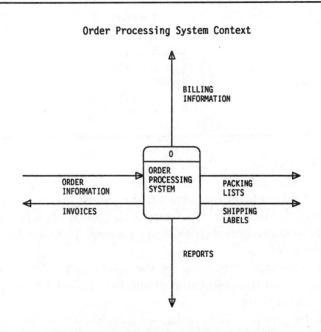

Order Processing System Context

Ask the group where this information is coming from and going to. They may tell you that "order information" comes from *customers* and "billing information" goes to *Accounts Receivable*. The other information goes to the *Customers, Sales Manager*, and *Shipping Department*. These are the external entities because, although they interface with the system, they are not part of it. By adding these four external entities, you have a complete first-level data flow diagram:

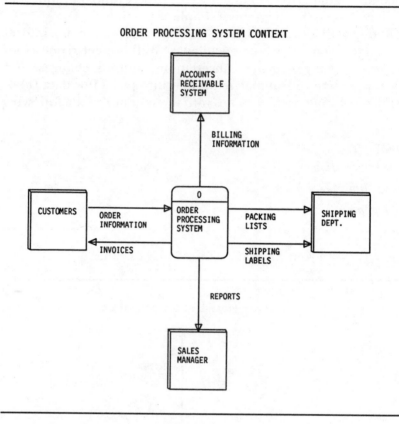

ORDER PROCESSING SYSTEM CONTEXT

2. Identify the Second Level of Work Flow

Keeping the first-level diagram on the board, move to the flip chart. Tell the group the next step is to define the second level of work flow. In other words, you are going to explode process *0—Order Processing System* into a second level.

Label the diagram at the top of the flip chart as *0—Order Processing System*. (The zero at the beginning of the title identifies the process from which it was exploded.)

Now, ask the group, "What processes does the Order Processing system include? What are the functions that you want it to do?" After much discussion on the many detailed tasks, they come up with three general ones:

enter orders

maintain customer file

print reports

These are the processes for your new diagram. This time, however, you will not use the board to build the diagram, for the following reasons:

- You usually do not have sufficient board space to draw another level *and* keep the previous level in place for reference.

- On all levels past the first, the diagrams go through many changes and interpretations. For example, one diagram might become cluttered with too many processes. Therefore, you might decide to combine four processes into one and allow another explosion to show the details. In this way, processes are combined, data flows are redefined, and parts of the diagram are shifted. As discussions progress, diagrams can change to such a degree that if you were using the board, you would be driven mad by having to constantly erase and relocate the shifting parts of the diagram.

So you need an easy way to move the parts of the diagram around. To do this, use a flip chart and those yellow sticky pieces of paper called Post-it® note pads, made by 3M. For simplicity, we will call them *notes*. As this second-level data flow diagram develops, the scribe writes the data flows, processes, data stores, and external entities on the notes while you place them on the flip chart. When a new process arises, the scribe makes a new note. When a location changes, you move the note to its new place.

Now that the group has identified the three processes for the second-level data flow diagram, you put these processes on notes and place them in the middle of the chart. (At this point, you have no idea what other processes might be added or where they will finally end up.) The flip chart looks like this:

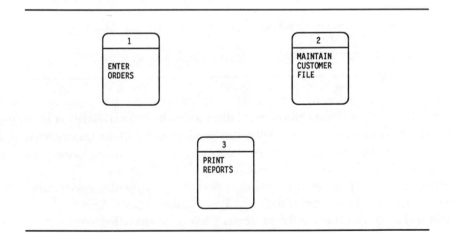

Ask the same questions as you did for the first level. "What data comes into these three processes and what data goes out? And where does the data go from here?" Several data flows and the data store *Customer File* are identified. The complete second-level diagram looks like this:

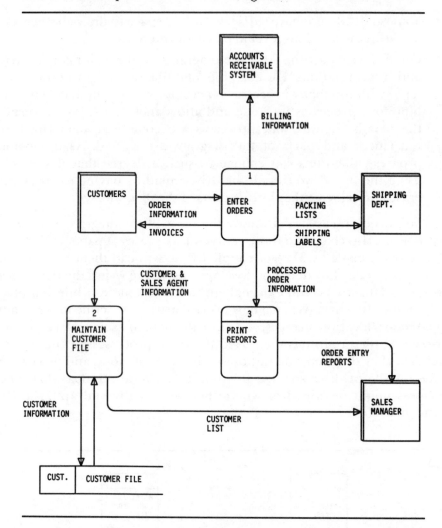

Two levels may be all you can accomplish in the first meeting. That is plenty. Do not keep the group more than an hour or two as this is mentally taxing work. People become very involved. You want to leave them at a point where they feel they really accomplished something rather than in the unresolved depths of a third-level data flow diagram.

After the meeting, you translate everything from the board and flip chart into two data flow diagrams. You either use a CASE tool or draw them by hand. Put them each on separate pages and title them:

Order Processing System Context

0—Order Processing System

Then copy these diagrams onto overhead transparencies.

The next day, hold a second meeting with the same group. Begin by reviewing what was done in the previous meeting. Using an overhead projector, show the two levels. Note any additional changes on the transparencies.

3. Identify the Third Level of Work Flow

The next step is to explode from the second-level diagram into several third-level diagrams. A separate diagram is created for each process. So the third level includes three diagrams named after the processes they exploded from:

1—Enter Orders

2—Maintain Customer File

3—Print Reports

Tack the original flip chart diagram to the wall for reference. On a blank flip chart page, write the title for today's first diagram:

1—ENTER ORDERS

Review the process called *1—Enter Orders*. In the same way as before, ask the group to break down this larger process. Perhaps they identify six subprocesses:

1.1—Add Orders

1.2—Change Orders

1.3—Print Orders

1.4—Print Packing List

1.5—Print Labels

1.6—Print Invoices

Put these processes in rectangles on notes and stick them to the flip chart. Identify the data flows, data stores, and external entities. In the previous level, notice the data flows coming in and out of the process. These must all be included in this diagram as well. In structured analysis and design, this is referred to as "balancing."

When this diagram is complete, continue with the second process, *Maintain Customer File.* Write the title on a blank flip chart page:

2—MAINTAIN CUSTOMER FILE

Break down this process into its subprocesses. They might be:

2.1—Update Customer File

2.2—Print Customer List

Put these processes on the flip chart and identify the data flows, data stores, and external entities. Is this beginning to sound repetitive? You bet! Once you get past the second level, a pattern develops where you identify, review, identify, review, and so on.

4. Continue the Levels

The number of levels you diagram depends on the size of the system and amount of detail needed for the specifications.

As before, guide the attendees through defining these levels. At the end, you have a series of data flow diagrams showing the work flow. You can take these diagrams into the JAD session as a starting point for discussion with the full group. The diagrams will probably change some more, but you will not be starting with a blank board. You can imagine how difficult managing this process would be in a full session with eighteen participants. Situations do arise, however, where you must define segments of work flow in the session. For example, sometimes new work flow is hinged on the outcome of open issues that are not resolved until the session. Nevertheless, you will benefit from having prepared these diagrams beforehand, in smaller, productive meetings.

Should You Document Existing or New Work Flow?

Depending on the scope of the project, you will define *existing* work flow or *new* work flow or both. Following are some scenarios.

Documenting Existing Work Flow Only

In some situations, the new work flow depends on other decisions that need to be made in the session. For example, you may have an open issue that asks, "Will data entry be centralized in one department or de-centralized among many?" The outcome of this issue directly affects many parts of the work flow. Therefore, you should cover existing work flow now, but avoid spending time defining a new work flow that depends upon agreements that have not yet been made. You may be tempted to define new work flow for both centralized and decentralized situations, but you are better off waiting until the session to design the new work flow.

Documenting New Work Flow Only

The most obvious scenario for defining new work flow only is when you are designing a new system that replaces no other. For example, a startup company that needs to develop all new accounting systems has no work flow in place to pattern after. This is the most challenging of all the scenarios because you are working in a near vacuum where you have, at best, sketchy specifications and varied opinions of how the system will work. More than ever, you need a structured way to pull these disparate visions together into an agreed-upon proposed work flow.

Another time to define only new work flow is when you are working under tight time constraints. Perhaps, due to a mandate from upper management, you have less time than you want to complete the JAD. Therefore, you must abbreviate parts of the preparation process. You may want to bypass defining the existing work flow and concentrate on the new work flow only.

Documenting Both New and Existing Work Flow

When none of these conditions apply, you can document both new and existing work flow. Begin by completely documenting the existing work flow as previously described. Then, walk the users through the existing flow again, making appropriate changes to reflect the new system design. In the end, you have both the existing and newly proposed work flow to bring into the session.

GATHERING PRELIMINARY SPECIFICATIONS

This part of the Research phase involves gathering information about the system requirements. This can include any preliminary specifications that will be further defined or reviewed in the session. For example, you will want to gather information on:

- data elements (or fields)
- screens
- reports

In the previous section, we talked about the importance of documenting work flow *before* the session because the task is too difficult in a full session with many people. You will find, however, that session agenda items such as data elements, screens, and reports can be more easily defined during the session. Therefore, you need not come into the session with complete prototypes. It does help, however, to have done some preliminary research in these areas.

What Information Do You Need?

The kind of information to gather depends on what you want to accomplish in the session. Do you need to define all the data elements for a new data base? Or will you be defining only a few new screens for an enhancement? Chapter 6 describes in detail how to document this information. For example, the "Data Elements" section shows exactly how the information you have gathered on data elements can be prepared before the session and used during the session. For now, the following summarizes what to gather during this research phase for data elements, screens, and reports.

Data Elements

For data elements, you might want a list of:

- *Existing data elements.* These are data elements used by the current system that also support the new system.
- *Changed data elements.* These are also used by the current system. However, their definitions must change for the new system.
- *New data elements.* These are data elements being defined for the first time.

Screens

For screens, you may want:

- *Screen flow.* This is a diagram showing how the screens branch.
- *Screen descriptions.* These describe the function of each screen.

- *Samples of existing screens*. These can be used in the session for reference only or as a basis for creating new screens.
- *Prototypes of new screens*. These are preliminary designs of how the new screens might look.
- *Screen messages*. These messages display on the screen and identify error conditions or confirm that an action has taken place (for example, a new order has been accepted).

Reports

Reports are similar to screens. You may want:

- *Report descriptions*. These describe the function of each report, including report name, general description, frequency of distribution, number of copies needed, distribution list, and sort specifications.
- *Samples of existing reports*. As with screens, existing report samples can be used in the session for reference only or as a basis for creating new reports.
- *Prototypes of new reports*. These are preliminary designs of how the new reports might look.

Where Do You Get This Preliminary Information?

You can get information on existing data elements, screens, and reports from MIS. Have the programmer/analyst prepare the data element list from the data dictionary and print sample screens and reports. For the reports, ask for samples with actual data that show typical output.

New prototype information comes from the users. You can meet with them before the session to develop prototypes for screens and reports.

THE JAD SESSION AGENDA

Although the session agenda has been evolving and probably will continue to change somewhat, now is the time to put it into as close to final form as you can.

The agenda is based on what you have learned from the following:

- preparing the Management Definition Guide
- doing familiarization interviews
- documenting the work flow
- gathering preliminary specifications

SESSION AGENDA

1. Overview
2. Assumptions
3. Work flow
4. Data elements
5. Screens
6. Reports
7. Open issues
8. Distribution list for the JAD Design Document

Figure 4-3 JAD session agenda

By now, you should have a good idea of what needs to be accomplished in the session. First, list everything that must be covered. Then, organize these items into a logical order. Figure 4–3 shows what a typical JAD agenda might look like.

Usually, the agenda items fall quite naturally into the order shown in Figure 4-3, or close to it. When they do not, however, you should design the agenda around the natural flow of the project, rather than force it into this order.

Examples of why you might want to change the sequence of the agenda items include:

- *Open issues.* Sometimes you can't even begin defining the system requirements without resolving at least some open issues. For example, for one project, we needed to determine whether the application would be a standalone system or be integrated into two existing systems. All the other agenda items would be affected by the out-

AGENDA FOR THE COMMISSIONS JAD SESSION

- Overview of existing Commissions system
- Define data elements
- Design new screens
- Review and modify existing reports
- Identify calculation routines
- Resolve open issues
- Determine distribution for the final document

Figure 4-4 Agenda for the Commissions JAD session

AGENDA FOR THE PROVNET JAD SESSION

- Overview of existing PROVNET system
- Review existing data elements
- Design new screens
- Determine screen access
- Revise agency status report
- Resolve open issues
- Determine distribution for the final document

Figure 4-5 Agenda for the PROVNET JAD session

come of this question. Critical issues such as this can be moved to the beginning of the agenda, just after "Assumptions."

- *Data elements.* Sometimes it makes more sense to identify the pieces of data the system will handle before getting into work flow. For example, one of our JAD projects involved defining user requirements for a new, complex insurance product. Before covering any part of the work flow, we needed to fully understand what information we were dealing with; that is, what were the data elements? Only then could we go on to define how the users would perform their jobs.

- *System-specific changes.* Sometimes the nature of the system being designed calls for a customized agenda. For example, we ran a JAD that did nothing but define transactions for a system. The agenda included three items: Assumptions, Transactions, and Open Issues. We ran another JAD whose purpose was to prioritize all the major data processing projects. The agenda included such items as: (1) review the project impact chart; (2) prioritize the projects; and (3) determine target dates.

Figures 4–4 and 4–5 show agendas for two projects. Figure 4–4 shows an agenda for modifying a Commissions system, and Figure 4–5 shows an agenda for modifying an online policy inquiry system.

Notice that the agenda items in these samples begin with verbs such as *define, design*, and *review*. These words further explain how each agenda item will be handled. If prototypes are available, you will probably be reviewing. If not, then you will be designing from scratch.

Having completed the Research phase, you are ready for Phase 3: Preparation.

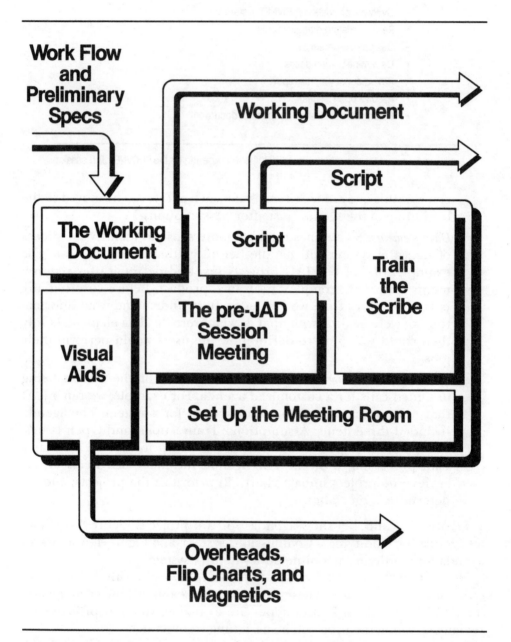

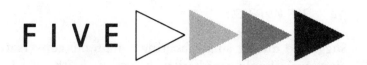

PHASE 3: PREPARATION

You have now interviewed users and MIS managers. You have documented work flow, gathered preliminary specifications, and prepared the agenda for the session. Having spoken with the people who know the system best, you should have a level of knowledge sufficient to talk intelligently about the system—at least at cocktail parties. Now, moving into phase three, you have pages and pages of *proposed* specifications which might include definitions of work flow, data elements, screens, and reports. Also, assumptions and open issues have been accumulating throughout the first two phases.

All this information needs to be compiled into a document that can be used in the session. Also, you need to prepare a script, train the scribe, create visual aids, hold a pre-JAD session meeting, and set up the meeting room the day before the session begins.

THE WORKING DOCUMENT

The Working Document is just that, something to work from in the session. It is a point of departure for defining the specifications. Although this document may have the look of a final copy, everything in the document is *proposed*. It contains the lists, diagrams, and text that people suggested during small meetings or phone conversations before the session. You should emphasize this point in the pre-JAD session meeting (discussed later in this chapter) so that people do not go into the session thinking that final decisions have already been made.

The Working Document is in the same format as the final JAD Design Document and can include:

- Title page
- Preface
- JAD overview
- Agenda
- Assumptions
- Detailed user requirements (including work flow, data elements, screens, and reports)
- Open issues
- Index

The following describes each of these parts.

Title Page

This includes the document title, system name, and date.

WORKING DOCUMENT
ORDER PROCESSING SYSTEM

June 20, 1990

Preface

This describes how the document fits into the overall JAD methodology.

PREFACE

This is a working document for the Order Processing system project. It includes proposed specifications to use as a starting point in the JAD session.

JAD Overview

This describes how the JAD methodology works. Unless there is a change in methodology, this overview remains the same for each document. Its purpose is to summarize the methodology for people who have not yet participated in a JAD. This overview is also placed in the final document because, even though the participants have, at that point, all been through a JAD, some of the nonparticipants receiving the final document may not be familiar with the methodology. This summary gives them the background they need to understand how JAD fits into the overall project.

The JAD overview includes:

- *Definition.* This is a one-paragraph description of the JAD methodology.
- *Benefits.* This summarizes how the methodology can help in systems design.
- *JAD Criteria.* This lists the characteristics a project should have to use the JAD methodology.
- *Open Issues and Assumptions.* Since they are integral to all JAD phases, the concepts of open issues and assumptions are described.
- *JAD Design Document.* This discusses the final document, its purpose, and the participants' role in creating it.
- *JAD Phases.* This summarizes what is involved in the five phases, emphasizing the roles of the participants. A chart shows each phase and its resulting documents.
- *JAD Roles.* This summarizes the responsibilities of the executive sponsor, leader, scribe, and full-time and on-call participants. For each role, the responsibilities are described for before, during, and after the session.

Our JAD overview spans five pages. It is shown on pages 301–305 in Appendix D, which is a sample final document for the project described in Chapter 11.

Agenda

This section provides information about the session. It includes the session agenda, list of session participants, and distribution for the final document. The following shows each of these parts.

Session Agenda

This describes what will be covered in the session.

SESSION AGENDA

The following items will be covered in the four-day JAD session to be held July 7 to 10, 1990.

1. Overview
2. Assumptions
3. Work flow
4. Data elements
5. Screens
6. Reports
7. Open issues
8. Distribution list for the final document

Session Participants

This tells who is attending the session, their departments, mail codes, and roles. The same list was included in the Management Definition Guide. You can copy that one and update it with any changes in participants. Also include the executive sponsor even though he or she may not be attending the session.

JAD Session Participants

Participants for the session include:

Name	Department	Mail Code	Role
Peg Barry	Accounts Receivable	2	Observer
Allison Brooke	Ordering	3	Team
Marko Chestnut	Law	3	On-call
Martha Clancy	MIS	7	Leader
Abby Eron	Ordering	3	Team
Michael Kowalski	Customer Service	4	Team
Linda Morgan	Shipping & Receiving	1	Team
Charles Mugler	MIS	9	Scribe
Ruth Noble	Ordering	3	Executive Sponsor
Anna Schwartz	Sales	4	Team
Jean Willis	Marketing	2	On-call
Sarah Wood	MIS	9	Team

Distribution for the Final Document

This lists all those who will receive the final JAD Design Document. This includes the participants and any others who should receive it. For a sample distribution list, see Appendix D (the final document).

Assumptions

The Assumptions section includes the sum total of business decisions agreed upon so far. These should be considered during the design process. For example:

ASSUMPTIONS

1. The system will handle order processing for the two acquisitions planned next year.

Detailed User Requirements

These sections of the Working Document tie directly to each particular agenda item covered in the session. The details of how to handle these agenda items and what the Working Document should contain are described in Chapter 6, which includes complete descriptions and actual Working Document samples. For now, the following lists them:

- work flow
- data elements
- screens
- reports

Open Issues

The Open Issues section includes the complete list to date of all unresolved questions that need to be answered either during the session or afterwards. For example:

OPEN ISSUES

1. How will we handle the different customer numbers that vary in size among divisions? For example, Milwaukee uses seven digits while Hoboken uses two letters followed by three digits.

Index

This is optional. It is recommended only if you have a text editor that automatically generates an index based on codes you enter. Otherwise, it is generally not worth the effort to prepare an index manually.

Notes for Reviewers

At the top of each section, you can make notes to the participants that tell them which parts of the Working Document to focus on in their pre-session review. For example, in the screens section of one of our documents, we included the following note:

For this chart, the participants from the Policy Issue department should review the first column, which shows the fields from the existing Policy Issue screens. Participants from the Variable Life department should review the second column, which shows the corresponding fields from the existing Variable Life screens. And everyone should review the third column, which shows the fields on the new screen.

Sending Out the Document

Send the document to all participants at least one week before the session begins. This gives them time to review the document and do any research or preparation necessary. If the timing works out, you can distribute it at the pre-JAD session meeting. Include a cover memo:

June 30, 1990

To: Order Processing JAD Participants

From: Martha Clancy

Subject: Working Document

    Attached is the working document for the Order Processing JAD session. This document, along with the Management Definition Guide distributed on June 16, will be the basis for our discussions in the session.

See you Monday at 8:30.

With the Working Document complete, you have accomplished a task that is as important as the session itself. The more clearly you can present these proposed specifications, the smoother the decision-making process will go when all the participants are together.

Now you are ready to continue with the other pre-session preparations.

PREPARING THE SCRIPT FOR THE SESSION

The script is your road map through the session. It tells you what to do and when to do it. While everything else produced in the session must make sense to a particular audience, the script must make sense to one person only—you, the JAD leader. So prepare the notes in whatever way is clear to you.

Unlike the script for a stage play, this script does not give the wording of exactly what to say. Instead, it guides you through each agenda item, reminding you of what you need to cover and which scribe forms and visual aids to use. In the session, you may refer to it frequently, but not as one would use notes at a podium when giving a speech. JAD sessions are not conducive to rigid word-by-word planning because you do not always know what will happen. You cannot predict how participants will respond to an agenda item. To return to the map analogy, the script tells you where the highways are, but it won't tell you when those highways are under construction or washed out in a flood.

What Do You Put in the Script?

The contents of the script mirror the JAD session, which is detailed in Chapter 6. In other words, the script shows how to handle each part of the session. It has three sections:

1. Introduction
2. Agenda Items
3. Notes

The following summarizes each section.

Introduction

This section addresses all the administrative items to cover before you get into the agenda. For example, for off-site sessions, it could include information about the rest rooms, making phone calls, and receiving phone messages.

This section also has notes on what to say about the session objectives and the general overview of the system being designed or enhanced. It lists what parts to review in the Management Definition Guide and any points you want to make when reviewing the agenda. For details on handling each of these items, see the "Opening the Session" section of Chapter 6.

```
Agenda Item:  Reports

Module:          New reports (p. 80 in the Working Document)

Action:          1.  Read new report descriptions.

                 2.  Revise these descriptions.

                 3.  Design new reports.  For each one:

                     a. Review the current report.  Is it
                        acceptable?  Does it need to be modi-
                        fied?  Use overhead transparencies to
                        make small changes.

                     b. Otherwise, use magnetics to design
                        new reports.

                 4.  Ask Michael if the new reports for his
                     department should be microfiched.

Visual aids:     Use transparencies for reports.  Use
                 magnetics for data element names to
                 build new reports.

Scribe forms:    Report description form
                 Report design form
```

Figure 5-1 Module in a JAD script

Agenda Items

This section walks you through the agenda. It is the main part of the script. Each agenda item (for example, Reports) is divided into modules (for example, New Reports). Then, for each module, the script describes:

- the module name and the corresponding reference page in the Working Document
- what actions to take
- what visual aids to use (for example, magnetics and overheads)
- what scribe forms are required

Figure 5-1 shows an example of a module for designing new reports.

For the Reports agenda item, in addition to "new reports," the script might also include a module for existing reports. The other agenda items are divided into modules in the same way.

Notes

Use this section for adding any comments that do not apply to a specific agenda item but that you want to keep in mind during the session. Figure 5-2 shows typical notes you might make.

```
Notes
  •  Summarize the agenda at the beginning and end of each
     day.

  •  When an agenda item is completed, put a check mark on
     the agenda flip chart.

  •  Remind Sarah to bring to the session a current hardcopy
     of the Customer Master File.

  •  At the break, make sure the coffee pot is full.

  •  Remember to prompt Anna about reporting needs from the
     Milwaukee division.

  •  At the end of each session, check with the facilities
     manager about table setup for the next day.
```

Figure 5-2 Notes from a JAD script

TRAINING THE SCRIBE

Training the scribe may be exaggerating the task. It is not exactly an Olympic event. But you do need to prepare the scribe for his or her role. This involves one meeting, at least a week before the session, where you:

1. *Summarize the role of the scribe.* The first question the scribe asks will probably be, "Why me?" and then, "Do I have to know how to take good notes?" Explain that the role does not involve note-taking in the traditional sense. Stress the importance of the role. You may be dealing with someone who is not pleased with being asked to scribe. Also, the person may not be aware of the critical impact he or she will have on the final product. This person was probably selected to scribe because of familiarity with the system and good communications skills. Explaining this may make him or her feel less encumbered with the task. The scribe's functions in the session are detailed in "The JAD Team and How to Select It" section of Chapter 3.

2. *Describe the JAD methodology.* If the scribe has not been in a JAD, walk him or her through the five phases, emphasizing how the scribe's role fits into the process.

3. *Discuss the project.* Because the scribe is familiar with the system, he or she will also know about the project to some extent. This is a good opportunity to get the scribe's comments. Make sure the scribe has reviewed the Management Definition Guide.

4. *Describe the session.* Show the scribe the printed agenda and discuss how each item will be handled. This is where you get into detail. Show what visual aids will be used. Point out when to make revisions directly in the Working Document and when to use specific scribe forms. Explain the task of making magnetics as new data elements or screen names arise. Give the scribe samples of all the scribe forms and explain each one.

In closing, ask the scribe to arrive at the session early to help set up the visual aids and scribe forms. Make plans to meet shortly after the session to review what was documented that day. And, of course, show appreciation for his or her help. You will be relying on the scribe a great deal throughout the session.

Daily Session Meetings with the Scribe

After each day of the session, you should meet with the scribe. Review all completed scribe forms. Are they understandable? (If they do not make sense now, they will have absolutely no meaning in three days.) Has all the information from the board been documented accurately? (Never count on the board information remaining overnight. The janitors may erase it.) Finally, are all the forms legible? (If you cannot read the scribe's writing, you need to find someone else for the job.)

VISUAL AIDS

This is the part of the process where you can roll up your sleeves and be creative. You can draw pictures on flip charts, write words in colored ink on magnetic shapes, and prepare overhead images that fill half the wall. The latent artist emerges here.

Visual aids help keep the participants focused and can clarify the decisions being made. You can build a picture of the application, step by step, as it evolves.

Flip Charts

Flip charts are used for information that you want displayed throughout the session. Good candidates for flip charts are:

- session agenda
- management objectives

- graphics for the system overview
- open issues

The following describes each of these flip charts.

Session Agenda

On a full flip chart page, list the agenda items. Throughout the session, you can refer to this list so people will always know where you are in the agenda. Since you must accomplish a lot of work in a short time, it is important to display the agenda to help keep the discussion on track.

When new subjects arise that you had not planned for in the script, you can note them on the agenda flip chart to assure they will not be overlooked. For example, someone may want to discuss the test plan. You can add this to the agenda flip chart, just before open issues.

Management Objectives

These are the objectives included in the Management Definition Guide. Since they describe what management expects to gain from the system, these objectives tie into the entire agenda. Having them on a flip chart helps when you review them at the beginning of the session and when you refer to them throughout the agenda.

Graphics for the System Overview

Near the beginning of the session, you or someone in the group will present an overview of the system. It's often said a picture is worth a thousand words. For a JAD session, you could say that a flip chart is worth, well, maybe 500 words.

Whatever concepts you want to communicate in this overview, you will probably want to come back to throughout the session. Take the time to show these concepts graphically.

People's familiarity with the system will range from those who are very involved in the details to those who have virtually no experience with it. It takes time to assimilate new concepts. The flip charts will help. For an example of a systems overview chart, see Figure 6–1.

Open Issues

As described in the "Open Issues and Assumptions" section of Chapter 3, open issues arise throughout the session. You need a way to note them and refer to them as the session progresses, and to review them on the last day. Therefore, write "Open Issues" at the top of a flip chart page. As issues arise, note them on the flip chart and number them sequentially. By the end of the session, you may have several pages of issues.

Magnetics

Now is the time we finally talk about these things called *magnetics*. They are thin sheets of vinyl cut in various shapes, sizes, and colors. Magnetic material, which comes in a long coiled flat roll about 1/2 inch wide, is cut and attached to the back of the vinyl shapes. This is what makes the magnetics stick to the board. You can then move the shapes around and place them anywhere you want. With special pens, you can write on the magnetics and remove the ink later with water (unless you mistakenly use pens that are not water soluble, in which case your magnetics are dedicated forever to whatever words were written). Suppliers for these magnetic materials are difficult to find. One such supplier is:

Ryan Screen Printing, Inc.
5412 West Burnham Street
Milwaukee, WI 53219

Figure 5–3 shows some typical magnetic shapes we have used.

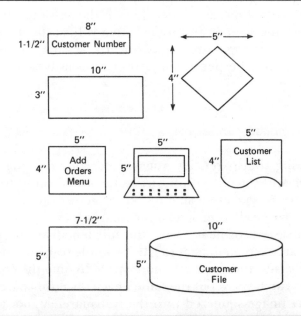

Figure 5-3 Sample magnetic shapes

The most common use of magnetics is for data elements. You can put data element names on magnetics that are 1–1/2 inches by 8 inches. Different colored shapes can be used to distinguish existing, changed, and new data elements.

One comment on preparing magnetics: Make sure you print large enough for everyone to see clearly. This may involve abbreviating somewhat. For example, instead of trying to fit "Increasing Death Benefit Rider" on one magnetic, you could print "Inc Death Ben Rider."

Overhead Projection

Most of you have probably used overhead projection in one way or another or have been to a meeting where it is used. It involves copying images of presentation material, not onto blank paper, but onto clear sheets called transparencies, foils, or overheads. Then, these transparencies are projected onto a screen or wall using an overhead projector. With special pens, you can write on the transparencies in various colors. You can erase these ink lines with moistened paper towels. This is not a particularly graceful process if you are trying to delete a tight area of small lettering. But it does work for simple edits.

In JAD sessions, transparencies work well for any situation where you are presenting something for *review*. In other words, you expect that most of what you display on the transparencies will remain, with only a few parts needing change. For example, you can use transparencies to show the proposed work flow developed before the session. Or you can use them for reviewing screens and reports that will, at the most, have minor changes. As these changes are decided in the session, mark them on the transparencies with the pens. The group can easily see the results displayed.

Some Comments on Using Transparencies

When making transparencies, you should copy the image as large as possible while still staying within the frame of the transparency. In other words, don't try to squeeze a six-line report across the top of the narrow side of a transparency. Instead, turn the image sideways, enlarge it (if your copier permits), and print it across the full length of the paper. In the same way, enlarge screen images to fill the whole transparency. The group can see more easily and you have more space to note the changes. Figure 5-4 illustrates two ways to put the same image on a transparency. The first one shows the image squeezed into the transparency (not a good use of

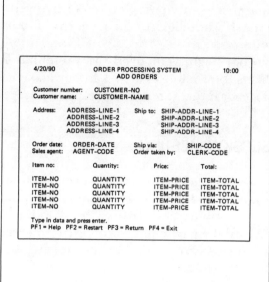

Figure 5-4 Putting images on transparencies

space), while the second one shows the same image turned sideways and enlarged (much easier to see).

If you are showing many overheads, tape a yellow transparency on the flat bed of the overhead projector where the light comes through. Then, place your transparencies for viewing on top of that yellow transparency. This softens the harsh white light of the projector and is much easier on the eyes of your viewers.

And finally, do not begin your overhead presentation with the statement, "I know this is too small to see . . ." Either make it large enough to see in the first place, or if that is not possible, do not use it at all.

THE PRE-JAD SESSION MEETING

This kickoff meeting is held at least a week before the session. The purpose is to establish management commitment, summarize the JAD methodology, and distribute and discuss the Working Document. Also, this is the first time all the participants will be together and have a chance to establish group identity. For that alone, the meeting is worthwhile.

What Happens in the Meeting?

The following describes what to cover in the meeting.

1. Establish Management Commitment

Introduce the executive sponsor. In a five- to ten-minute presentation, he or she should summarize the objectives of the project and how it will benefit the company. Perhaps this person can talk about how the project came to be and how it will solve certain problems. He or she can describe the time and resources that have been committed to the project. And the executive sponsor can emphasize how the JAD methodology, particularly the session, is a key part of the whole development effort. "People are relying on the session participants to define the system requirements."

2. Summarize the JAD Methodology

Now the agenda turns back to you. Discuss how the JAD methodology is being used to support this project. Show how it fits into the systems development life cycle. Review the five JAD phases, highlighting the participants' responsibilities in each phase. This discussion might not be required for an experienced group who has been through JADs before. For example, we do many JADs for new insurance product introductions that involve the same application systems and therefore the same people. Participants who have recently endured a five-day session do not need to hear that "the session" is Phase 4 of the JAD methodology.

3. Distribute the Working Document

Explain with emphasis that everything in the document is *proposed*. You might say:

Although this Working Document appears to be in final form, it is only a starting point for the session. Everything in this document is *proposed*. It includes specifications that have come from you.

For each section, we may be simply modifying the specifications or we may be starting from scratch. When we cover screens, for example, you may find that the proposed screen designs are close to what you need. In that case, we can just review them and change some fields. Or, you may find that the screens have nothing to do with what you require. In that case, we can start at the beginning, with a blank board and new ideas. It's up to you.

So, as you review the document before the session, realize that much of it will change. Make your comments directly on the pages so we can talk about them in the session.

Have the participants turn to the table of contents as you summarize what is contained in the document. Do not rush through it. Even though you are intimately familiar with the document, no one else has even seen it. So avoid breezing through in a whirl of details.

Highlight the sections you want the participants to spend the most time reviewing. Only the most conscientious (or those who also enjoy reading software manuals) will read every word of the document. So show them what areas to concentrate on. Make their review as productive as possible. For example, you might say:

Spend time reviewing the data elements. In the session, be prepared to confirm or change them. For the existing data elements, are these definitions consistent with the mainframe data dictionary?

Remind the participants that the comments included at the top of each section describe how to review that section and how it will be handled in the session.

SETTING UP THE MEETING ROOM

It is now the day before the session. The Working Document has been prepared and distributed. The visual aids are ready. You have several rolled-up flip chart pages and a JAD supplies box filled with magnetics, overheads, scribe forms, board pens, and so on. Now, all you have to do is set up the room to accommodate all these items as well as all the session participants. This is where you can set aside your mental efforts for some physical activities: moving chairs, adjusting tables, setting up the over-head projector, dragging flip chart stands across the room, and lugging a few boxes around.

Arrange to have access to the room on the day before the session. If possible, avoid setting up the day of the session. These activities are a burden you do not need at a time when you have the script for a five-day workshop rolling around in your head. Running back to the office ten minutes before the session to get the overhead transparencies you left next to the copying machine does not add to your composure as you open the session.

To set up the room:

1. *Arrange the tables in a hollow square.* In other words, the tables should create a complete square with an open space in the middle. Break the square toward the front of the room (by leaving one table out) so that you can go inside to project the overhead transparencies. You, the JAD leader, along with the scribe, are the only people on the front part of the square. Any participants sitting there would not be able to see the board. We usually use two large tables for the leader and the scribe and a small table or a couple of chairs off to the side for the box of supplies and extra scribe forms. Figure 5–5 shows how tables are arranged for a typical session.

   Facilities management will usually set up the tables for you. But more often than not, you will need to make adjustments.

2. *Hang the flip charts.* Place the "Session Agenda" flip chart in a visible place toward the front of the room. You will refer to it often. Other flip charts (for example, graphics for the system overview) can be hung along the sides. And finally, "Open Issues" should be in a front corner, in an easily accessible spot. While the other flip charts are taped on the walls or boards, "Open Issues" remains on a regular

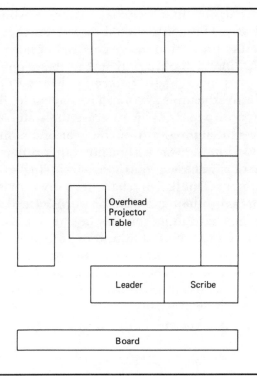

Figure 5-5 Table arrangement for a JAD session

flip chart tablet placed on a stand. Have several pages ready because open issues arise continually throughout the session.

3. *Check the things that can go wrong.* This refers to the typical, but often overlooked, *mechanical* things that can go awry in any kind of session. This includes checking items such as overhead projector bulbs and board pens that may have run dry since the last session. All this may sound trivial and perhaps overly obvious, but these are just the kinds of glitches that can throw off your rhythm. It's worth a test run.

Set up the overhead projector. Check both the in-place bulb and the spare. Project the image and focus it. Walk around the room. Will everyone be able to see? If you are projecting from the middle of the room, which seat will be in the path of the light? You will need to ask that participant to relocate while you are projecting transparencies, or that person will be jolted by a beam of blinding light. Make sure that the surface you are projecting upon does not produce a glare that has that high-beams-in-your-eyes effect. Most white boards do not work well for this reason. Instead, use a projection screen, light-colored wall, or boards with a photo-gray projection surface.

Know where the light switches are. Test which lights can be dimmed, if any. The lights should be low enough for your participants to see the projection image clearly, but not so low as to create the kind of "mood lighting" that has them nodding out by the second transparency. Finally, determine where to place the "props" you need for overhead projection. These include the transparencies, pens, water, and paper towels to erase the water-soluble pen lines. Keep lightweight items away from the rear of the projector where the fan blows. Otherwise, you will begin your presentation with a dramatic storm of windblown transparencies flying across the room.

Check all pens. Do the flip chart pens work? Will the board pens carry on for more than a couple of words? Do you have plenty of them along the board rail? (You may be one of those with the strange habit of carrying the pens away and depositing them in other parts of the room. This results in your dashing about searching for pens whenever you need to write something on the board.) And, if the boards require a liquid solution to erase, do you have sufficient "erasure juice"? Trivial but essential.

A note on pens: You have flip chart pens, board pens, and pens

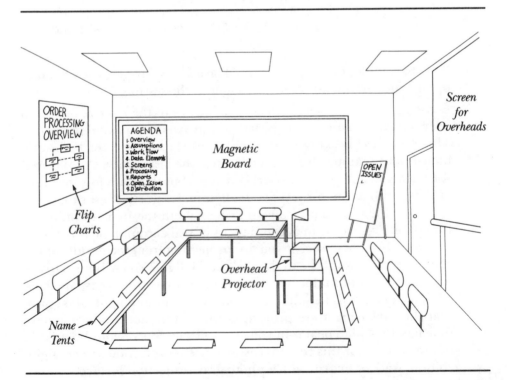

Figure 5-6 JAD session meeting room

for overheads and magnetics. Be sure to keep them straight. For example, using anything but the correct water-soluble pens on the magnetics will create permanent images. And if you use a flip chart pen on the board, make sure you write something worthwhile, because it will be there for good.

4. *Distribute name tents and pens.* Place the blank name tents and some marker pens around the tables so the participants can fill them in when they arrive. No preplanned seating arrangement is required. One interesting thing you will notice: More often than not, people will take the same seat every day of the session. The habits learned in grade school are with us all our lives.

Figure 5–6 shows how the meeting room can be set up.

Having compiled the Working Document, prepared the script for the session, trained the scribe, created the visual aids, held the pre-JAD session meeting, and set up the meeting room, you are ready for Phase 4: The JAD Session.

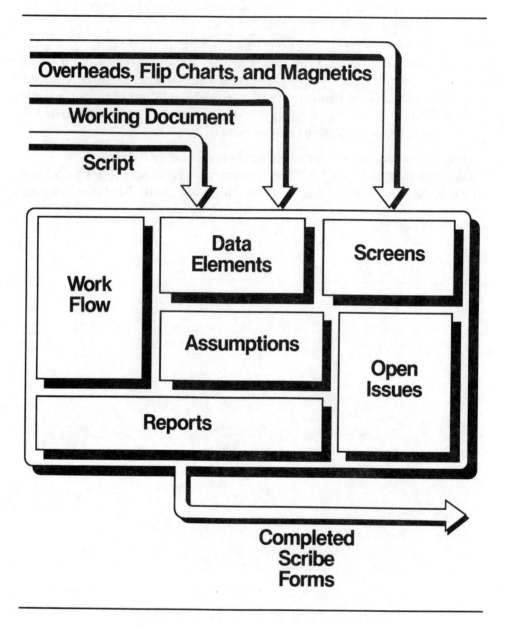

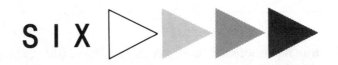
PHASE 4: THE JAD SESSION

In the JAD methodology, the session is the main event, "the moment you've been waiting for." Up to this point, all efforts have been in preparing for the session. Now, all the information gained in interviews, compiled in the Working Document, and illustrated in visual aids, comes together to support this three- to five-day structured workshop. The typical JAD session will cover the following agenda:

- assumptions
- work flow
- data elements
- screens
- reports
- open issues

In the session, you will use the Working Document as a basis for defining the final specifications. As you embark on each agenda item, different levels of detail are available from the document based on how much pre-session "homework" you have done. For example, in the Screens section, you could find yourself in any one of the following situations:

- You have already gathered proposed specifications for new screens and now the group need only review them and make some minor changes.

- You have gathered proposed specifications (as in the previous example), but in the session, the group agrees that these screens do not meet the needs of the new work flow revised earlier in the session. In other words, you must start over.

- You come into the session with few or no screen specifications at all. The only thing certain is that screens must be designed.

Whether you have comprehensive specifications or none at all, the descriptions in this chapter apply. For each agenda item description, you will see two parts. The part *before the session* shows all the specifications you could possibly have gathered coming into the session. The part *during the session* describes how to define specifications from scratch as if you came in with nothing at all. This way, you can tailor each agenda item to the level of preparation you have done. Also, if the participants decide to abandon the specifications proposed in the Working Document (leaving you with a large blank board and four colored pens), you are prepared to start at the beginning for any item on the agenda.

OPENING THE SESSION

The first two minutes of any session can be the most tense. Leading a session is in part a performance, like teaching a class. It's normal for the JAD leader to have "opening-night" butterflies. Anyone who has had a speaking part in their fifth-grade school play knows the feeling. Vice Presidents are there. Your boss's boss is there. And users and MIS people are there, sitting in close proximity.

Everyone has their eyes on you. Can you pull it off? You close the door, clear your throat, and say, "Welcome. . . ."

1. Administrative Items

Begin by reviewing the administrative items that answer questions about how the meeting will run. This includes:

- *Schedule.* Tell what time sessions begin and end, and when the breaks occur.

- *Restrooms*. Tell where they are located.
- *Phone calls*. Provide the phone number where people can leave messages (no calls directly into the room). Tell them where they can make calls during breaks.
- *Introductions*. Introduce those people with particular roles, such as the executive sponsor (if he or she is attending), the scribe (the one with the twelve sharpened pencils), and the observers (the ones sworn to silence). Ask the participants to fill out the name tents that you have placed on the tables. These should suffice for other introductions. Also, refer them to the Working Document, which lists the partici-pants and their departments.

2. Session Objectives

Describe what you expect to accomplish in the session. You might say:

We are here in this session to define your requirements for order processing. This new system will replace the existing order processing systems (both manual and automated) across all divisions of the company. We will use the Working Docu-ment (distributed last week) as a basis for each item on the agenda.

As decisions are made, we will slow down the pace while those agreements are recorded by Charles, our scribe. He will read them back to confirm the accuracy. Then, all this information goes into a final document which will be used to build the final application.

---

3. System Overview

You or someone in the group now presents an overview of the system. This high-level summary answers such questions as: Which departments are affected by the system? Why is it being designed or enhanced in the first place? What problems are you dealing with? Figure 6–1 shows a diagram that could be used for the system overview in the Order Process-ing JAD.

Also in this overview, you can clarify the terminology that will be used throughout the session. Remember that people's familiarity with the sys-tem under review may vary tremendously. Keep this overview brief and as free of technical jargon as possible.

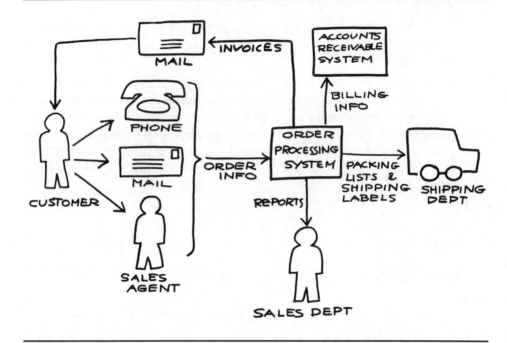

Figure 6-1 System overview flip chart

4. The Session Agenda

Walk through the session agenda. Refer the group to the flip chart, elaborating on how each agenda item will be handled.

5. Management Definition Guide

Although the participants received this document before the session, reviewing the main points helps reinforce why the system is being designed. Read aloud (verbatim or in paraphrase) the purpose, scope, and objectives sections. You may have prepared a flip chart of these objectives which you can refer to now and throughout the session.

Now, you are ready to delve into the Working Document and start making those decisions we have been alluding to. This begins with a review of the assumptions.

ASSUMPTIONS

Assumptions are entwined with open issues. It is difficult to talk about one without the other. The following shows how they both can evolve throughout a JAD project.

Phase 1: Project Definition	During an interview with a user manager, an issue arises concerning who will maintain the Customer File. Will changes be keyed in separately from different divisions as it is done now, or will home office staff handle all data entry? This goes into the Management Definition Guide as open issue number one:

OPEN ISSUES

   1. Will the Customer File be maintained by each division or by the home office staff?

Phase 2: Research	Separate groups of users discuss the question but cannot resolve it without other users present.
Phase 3: Preparation	This issue is copied from the Management Definition Guide (along with the other issues) into the Working Document.
Phase 4: The JAD Session	During the session, the team discusses another question regarding whether the system should be able to handle order processing for companies acquired in the future. The team agrees that it should. The leader prompts the scribe to document this assumption on an Assumption scribe form:

ASSUMPTIONS

   1. The system will handle order processing for the two acquisitions planned next year.

At the end of the session, the team returns to open issue number one (maintaining the Customer File). They agree on a solution that works for everyone. Again, the leader prompts the scribe to document this new assumption:

ASSUMPTIONS

2. All departments will send customer file information to the home office on specially prepared coding sheets. This information will then be entered from a centralized location.

Phase 5:
The Final Document

After the session, all the assumptions come together in the final document in a section called "Assumptions."

This example has shown the development of two assumptions. An actual session can produce many more assumptions (as well as open issues). Although the range varies tremendously, the average for us has been 15 assumptions per session.

The length of time spent on assumptions depends, of course, on how many assumptions you bring into the session. If an assumption turns into an open issue (or if some issues must be addressed at the beginning of the session), much more time may be required. In such a situation, the participants may become skeptical about the session's progress when the first day ends with them having covered only three pages of a one hundred-page document. Assure them that this is not uncommon. These important assumptions are the basis for defining the rest of the requirements.

Before the Session

Assumptions have been accumulating since the JAD project began. The Working Document lists them all, including the assumptions from the Management Definition Guide (which you copied directly into the Working Document) and those that may have surfaced after that.

During the Session

Again, remind the participants that all these assumptions are for review—they can be changed. Read each assumption to the group, then open it up for discussion. Each assumption will either:

- stay as it is (if everyone agrees with the wording)

- be revised
- become an open issue (if group consensus cannot been reached)

The scribe documents the minor changes directly on the pages of the Working Document. He or she documents new assumptions on the Assumptions scribe form and reads them back to the group. These new assumptions will continue to arise throughout the session.

WORK FLOW

Work flow shows how information moves through the system. This can include work flow from both user and system perspectives. Usually, the flow is analyzed from the user's viewpoint. Therefore, typical specifications are in terms of business activities.

Before the Session

During user interviews, work flow was identified. In small meetings, the leader guided a few key users through the process of defining work flow for the existing system as well as proposed work flow for the new system. Data flow diagrams or other techniques were used to document this flow. These diagrams were put into the Working Document as well as onto overhead transparencies to use in the session.

During the Session

Now, with everyone present, the leader walks through the *existing* work flow using an overhead projector. Participants may have comments or minor changes on this flow, but generally there is not much discussion about what already exists.

In contrast, *new* work flow can generate lively debate. Here you are talking about changes that will directly affect the participants' working environments and daily procedures. Will they need to hire new staff, decrease existing staff (now that these functions are automated), or completely reorganize the department? There is much at stake.

Present the new work flow one step at a time (or in data flow diagrams, one level at a time). Do not imply that any of the flow has been finalized. Use phrases like, "This is how the procedure *could* work . . ." or "One approach is . . ." After all, many participants have not yet been involved. They need to establish ownership in the concept by seeing their ideas included in the diagrams.

As participants agree on changes, note them on the transparencies so everyone can clearly see the decisions. Your diagram could end up looking something like this:

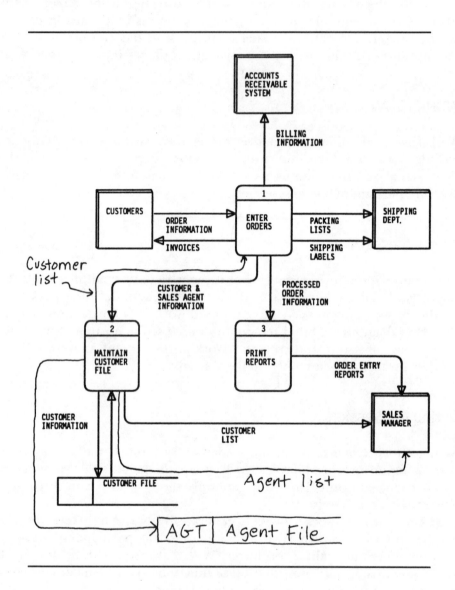

The scribe notes these changes in the Working Document.

After this first walkthrough, move through the new work flow one more time, reviewing for accuracy and consistency. Changes made in one area may now affect another.

Of all the agenda items, work flow is the most important to define upfront, before the session, in small meetings. Although the diagrams will change in the main session, the process is much smoother when you have a basis from which to start. If you have not yet done this, use the same process described in the "Documenting Work Flow" section of Chapter 4.

DATA ELEMENTS

Every piece of information that will be entered, processed, stored, displayed, and reported by the system is packaged into units called data elements (or fields). These are the building blocks the group will use throughout the session to design screens and reports and build the data dictionary.

Before the Session

Proposed data elements were defined and listed in the Working Document. For example:

DATA ELEMENT DESCRIPTIONS

Name:	Customer Number
Length:	7
Format:	Numeric
Description:	A unique number assigned to each customer.

Name:	Customer Last Name
Length:	20
Format:	Alphanumeric
Description:	The last name of the person ordering the item.

These definitions can include other information such as allowable values (for example, the allowable values for Customer Number could be 1000000 to 5999999). Also, companies that program in the COBOL language can do the following:

- For "Name," use the COBOL name.
- Combine "Length" and "Format" into one item called "COBOL format."

For example, the data element descriptions shown previously would look like this:

DATA ELEMENT DESCRIPTIONS

Name: CUST-NO

COBOL
Format: (9)7

Description: A unique number assigned to each customer. Allowable
                 values are 1000000 to 5999999.

Name: CUST-LAST-NAME

COBOL
Format: X(20)

Description: The last name of the person ordering the item.

However, make sure the users understand this terminology and are comfortable with it before introducing it into the session. If there is any doubt, do not use the COBOL format.

In the Working Document, data elements are organized into three groups:

- *Existing data elements*. These are current data elements that will be used by the system and already exist somewhere in a data base or file. Even though these data elements will not change, review them to familiarize the group with their definitions.

- *Changed data elements*. These are current data elements that can be used in the new system by changing the definition or range of values. For example, the existing data element called Customer Region might have two new codes added for the two new additional regions.

- *New data elements*. These are data elements proposed for the new system. Their definitions include data element name, length, and a short description (as previously shown).

The system you are designing could involve data elements from any of these three groups. If you are enhancing a system, for example, you might use all existing data elements with changes to just a few. On the other hand, if you are designing an automated system to replace the manual one, then you need to define all the new data elements.

The required data elements have been written on magnetics. Different colors (of the magnetics or of the ink used to write the names) indicate which group the data elements are from. For example, white magnetics could be for existing data elements, yellow for changed, and blue for new.

During the Session

All the magnetics are arranged in columns on the board. To find them easily, sort them first by color (existing, changed, or new), then alphabetically. Figure 6–2 shows what the board might look like.

Review each data element for its correct name, length, and definition.

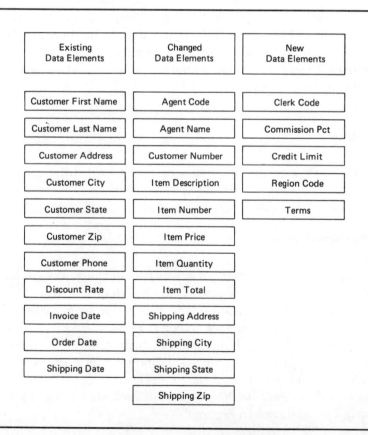

Figure 6-2 Magnetics showing data elements

New data elements will arise and others will be removed. For example, online systems today can use the log on ID code to determine who is accessing a particular application. Therefore, it may not be necessary for the clerk to enter a separate code to identify who is entering the data. Consequently, the separate data element, Clerk Code, may no longer be needed. These kinds of changes occur throughout the session.

As new data elements are defined, the scribe makes a magnetic to add to the list and documents the definition on a Data Element Description scribe form. As data elements are deleted, the scribe pulls them from the board to keep in a separate pile in case they are called back. (One data element could be removed and reinstated three or four times before the participants make up their minds.)

Now, the magnetics on the board display a complete list (to this point) of data elements required by the new system. These data elements will be used to build the screens and reports later on.

SCREENS

This part of the session involves defining how users enter information into the system. These days, this is usually done through data entry screens. There are two parts to defining screens: you need to define *screen flow* (how users branch from one screen to the next) and *screen design* (what actually displays on each screen).

Screen Flow

Screen flow uses a series of menus or other branching techniques to define how users access the various system functions.

Before the Session

Based on interviews with users, a *screen flow diagram* was prepared and included in the Working Document. In this diagram, you documented how users branch through menus and submenus to get from screen to screen. Figure 6–3 shows a sample screen flow diagram.

You prepared a 5-by-7-inch magnetic of each main menu selection (the boxed items in Figure 6–3) and a smaller magnetic of each submenu selection (the numbered items in the same figure).

If you needed to describe the purpose of each screen, then you created *screen descriptions* as shown in Figure 6–4.

Screen Flow Diagram

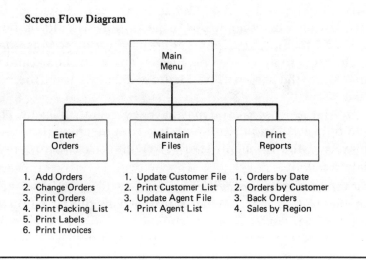

Figure 6-3 Screen flow diagram

During the Session

Now, let's take a look at how to walk the participants through the process of defining screen flow. Remember, we are assuming you have done no preparation and that the Working Document has nothing on screen flow. This is to prepare you for any situation that might arise requiring you to work from scratch. To define screen flow, follow the steps on page 108.

SCREEN DESCRIPTIONS

Name:	Add Orders
Function:	Allows users to enter orders for customers that are already in the Customer File.

Name:	Change Orders
Description:	Allows users to make changes to orders that have already been entered.

Name:	Print Orders
Description:	Prints order forms for all unfilled orders including back orders.

Figure 6-4 Screen description

1. *Identify the main menu selections.* This produces the items shown in the boxes in Figure 6–3. Ask, "What are the *main* functions the screens will handle? In other words, what options should be on the main menu?" As the participants identify these functions, make 5-by-7-inch magnetics for each one. Place them on the board and draw lines that show the screen flow. Figure 6–5 shows what the board looks like.

   For larger systems, you may have several menu levels. This is when the phrase *sufficient board space* takes on meaning. For example, you may find the submenu levels stretching clear across two or three board panels.

2. *Identify submenu selections.* This produces the numbered items shown in Figure 6–3. Ask, "What is involved in the first function called *Enter Orders?* What options should that submenu contain?" Write these functions on smaller magnetics and place them below the menus. When all functions are identified, the board is filled with magnetics, as shown in Figure 6–6.

   Putting this information on magnetics works much better than just writing it on the board, because with magnetics you can:

   • make changes more easily by simply moving the magnetics as requested. This is especially helpful when dealing with complex screen flows where users are moving back and forth among screens based on the values they entered into those screens.

   • view the entire screen flow. When the screen flow is complete, you can move the magnetics in their final arrangement over to a corner of the board. Then, as you design the screens, the overall screen flow will be available for reference.

3. *Describe the screens.* This produces the list shown in Figure 6–4. Have the participants finalize the screen names and create descriptions of the functions. The scribe notes the screen names and descriptions on a Screen Description form.

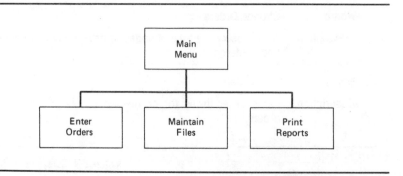

Figure 6-5 Magnetics showing menu screen flow

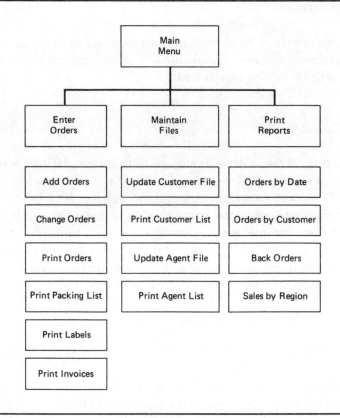

Figure 6-6 Magnetics showing complete screen flow

4. *Review the complete screen flow.* When the group feels the screen flow is complete, ask, "What tasks have been overlooked in these screens? Can any of the screen functions be combined? Are the screen names simple and clear, accurately reflecting their purpose? Will the data entry clerk know exactly what these names mean?"

Now, with screen flow defined and documented by the scribe on a Screen Flow Diagram scribe form, you are ready to design the actual screens.

Screen Design

If there is a high point in a JAD session, it is this portion of the agenda— *screen design.* Here, the group determines how the screens will actually look. This can be fast-paced and animated as the leader moves magnetics around the board while participants command what fields to add and relocate. Following is a description of how it works.

Before the Session

Interviews before the session can generate samples of existing screens and prototypes of new screens. The following describes how this information was put into the Working Document.

Samples of Existing Screens

Any existing screens that relate to the system being designed are included in the Working Document. In the session, these screens might be used as is or updated to reflect new enhancements.

We keep files (stored in the computer) of screens that frequently come up in JADs. For example, we save complete screen samples for the online policy inquiry system. When we need to bring certain screens into a session, we print them out, ask the appropriate project manager to review the screens to assure they are current, and add them to the Working Document.

Prototypes of New Screens

If you have pursued specifications with users to the point where they have worked up some prototype screens, then you have added them to the Working Document. You must be careful, however, not to be too ambitious in your preparation before the session for this part of the agenda. You do not want to spend a lot of time developing prototype screens if certain other specifications have not been addressed. For example, spending time on screens is fruitless without knowing what data elements are in the mix. On the psychological side, when you present these prototype screens, be careful not to give the impression with your beautiful hardcopy printouts that the final decisions have already been made. Emphasize again that everything in the Working Document is proposed.

For both existing and prototype screens, you have prepared overhead transparencies. Even if the existing screens are for reference only, you will be glad to have the transparencies on hand when the group decides to look at one in detail and perhaps tear it apart.

During the Session

If you have done all the preparation just described, then you can use an overhead projector to display the screens for the participants to review. (Refer the squinting participants to the copies in the Working Document.)

You are either modifying existing screens or reviewing prototypes of new screens. As you display each transparency, ask what changes need to be made. Mark the changes directly on the transparency.

If the changes become extensive or if you have done no preparation (as is often the case with screens), you need to begin with a blank board. You can plan on designing about three screens per half-day session. Follow these steps:

1. *Set up the board.* Draw two large empty frames (representing screens) on the central panels of the board. If you have enough board space, draw three or four frames. Arrange the magnetics of data elements (defined earlier) off to the side on a separate board panel.

2. *Define the headers and footers.* This is the information that displays at the top and bottom of every screen. It may already be determined by your company standards. Write the headers and footers directly on the board at the top and bottom of the screen frames, not on magnetics. This may include the system name, screen name, and perhaps the date and time. Figure 6–7 shows an example of what it might look like. These headers and footers remain on the board throughout the screen definition process.

3. *Design the menus.* Begin with the main menu and continue through the submenus. For simple menus, write the design directly on the

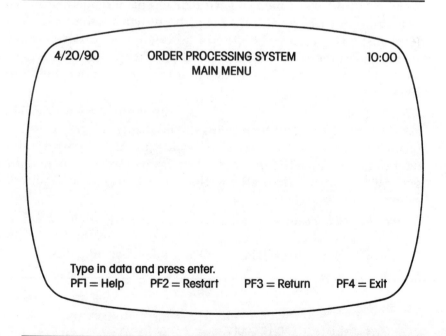

Figure 6-7 Headers and footers for screen design

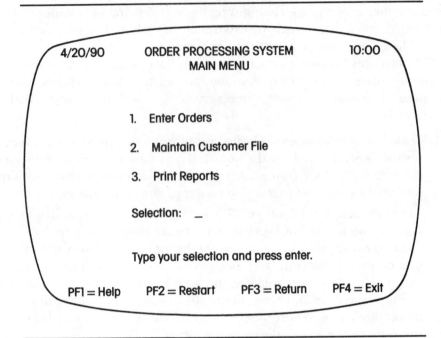

Figure 6-8 Main menu design

board. For complex menus, magnetics allow you to move the selections around, experimenting with different arrangements. While
defining a particular screen, keep the design of the next highest
level displayed on the board. For example, while designing submenus, keep the main menu in view. This keeps the screen flow in
context. Figure 6–8 shows what the main menu design might look
like.

4. *Select the fields for the screens.* Begin by adding the screen title. Then,
   ask the participants, "Which fields should display on the screen?" As
   participants call out the required fields, move the corresponding
   data element magnetics into the screen frame. Arrange them in their
   general locations. When all the fields are there, adjust their posi-
   tions.

5. *Define the field labels.* For each field, define the field labels that
   prompt users what to enter in that field. For example, the field
   CLERK-CODE might be labeled "Order taken by:" like this:

Order taken by: | CLERK-CODE

Write the field labels directly on the board to set them clearly apart from the fields.

6. *Fine-tune the placement of the fields and field labels.* Consider space limitations. Most screens are 80 characters wide and 24 lines long.

This is the point in the session where the group splits. There are those who say, "I don't care where you put the fields, let's get on with it." Others really get into the details, offering such comments as "Move that field up a little and to the right . . ." as if they were hanging a picture on the living room wall. But these comments are worthwhile. Good screen design results in screens that are easier to read and more efficient to use.

Now, within the frame drawn on the board, you have a completely designed screen with:

- headers, footers, and a screen title (written on the board)
- fields (shown on data element magnetics in their proper location)
- field labels (written on the board)

The scribe documents the screen design on the Screen Design form. Figure 6-9 shows the completed screen design. You can follow these six steps for all screen designs.

7. *Determine screen messages.* Identify all the messages that could display

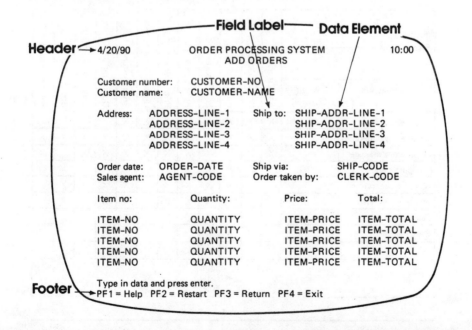

Figure 6-9 Completed screen design

Screen Messages

Screen name: _Add Orders_
Change Orders

Message type: Confirmation _____ Error _X_ Other _____

Conditions (When does this message display?)
When users enter an item number that
is not on the Inventory data base.

Message text: _THIS ITEM NUMBER DOES NOT EXIST._

Figure 6-10 Screen message description

on the screens. Then for each message, identify its type: Is it for confirmation (acknowledging that a certain action has been taken) or for errors (the data entered has not passed all the edits)? Next describe the conditions under which the message displays and how it

---

Screen Access by Job Function

Job Function	No. of Terminals	Add Orders	Change Orders	Print Orders	Print Packing List	Print Labels	Print Invoices	Update Customer File	Print Customer File	Orders by Date	Orders by Customer	Back Orders	Sales by Region
Order entry clerk		X	X	X	X	X	X	X	X				
Order entry supervisor		X	X	X	X	X	X	X	X	X	X	X	X
Order entry manager		X	X	X	X	X	X	X	X	X	X		X
Shipping clerk				X	X	X	X					X	
Shipping supervisor				X	X	X	X		X	X		X	
Shipping manager				X	X	X	X		X	X		X	
Sales agent									X				
Sales manager		X	X	X				X	X	X		X	X

I = Inquiry
U = Update
X = Either

Figure 6-11 Screen access by job function

will be worded. The scribe documents this information on the
Screen Messages scribe form as shown in Figure 6–10.

8. *Determine screen access.* If the application has security requirements,
now is a good time to define them. Let's say you want to define
which job functions can access which screens. On the board, create a
large chart. Across the top, list the screen names. (You can use the
already created magnetics for this.) Then, down the left side of the
chart, list the job functions. For each job, ask the group, "Which
screens can this job access?" Some screens are designed to be used
for either inquiry (users can read the screen but not add or change
data) or update (users can read *and* update information). In this case,
mark the space on the chart with either "I" (for inquiry) or "U" (for
update). If you do not need to make this distinction, just mark X's in
the columns. Figure 6–11 shows how this chart might look.

Now your screen design is complete. But before going on to reports,
let's explore another area of screen design.

Human Factors in Screen Design

When designing screens, you are dealing with a number of factors outside
your control: There are the screen standards that your company has set.
There are hardware and software limitations. And, of course, the screens
themselves must accomplish certain functions. There is also a whole area
of screen design that is critical, but inadequately considered—the area of
human factors. This is the aspect of screen design that deals with how users
interact with the screen. Much research has been done in this area. The
resulting guidelines can be applied to your screen design techniques.
Here are some human factor tips to consider when designing your next
batch of screens:

- *Uppercase versus lowercase.* You may have noticed that the screen
samples given in this book show field labels in lowercase letters.
This is because lowercase is easier to read. If this is true, then why
are most screens in uppercase letters? Years ago, computer systems
could handle only uppercase. But now, even though most systems
can handle both, many of us continue in the uppercase tradition
because "that's the way we always did it."

    To see the difference in clarity, compare the following two pas-
sages. They both have the same text. The only difference is that the
first one is all in uppercase letters, and the second has mixed upper-
and lowercase.

TO THE LOOKING-GLASS WORLD IT WAS ALICE THAT SAID
I'VE A SCEPTRE IN HAND I'VE A CROWN ON MY HEAD.
LET THE LOOKING-GLASS CREATURES, WHATEVER THEY BE
COME AND DINE WITH THE RED QUEEN, THE WHITE QUEEN
    AND ME.
THEN FILL UP THE GLASSES AS QUICK AS YOU CAN,
AND SPRINKLE THE TABLE WITH BUTTONS AND BRAN:
PUT CATS IN THE COFFEE, AND MICE IN THE TEA—
AND WELCOME QUEEN ALICE WITH THIRTY-TIMES-THREE!

            —LEWIS CARROLL

To the Looking-Glass world it was Alice that said
I've a sceptre in hand I've a crown on my head.
Let the Looking-Glass creatures, whatever they be
Come and dine with the Red Queen, the White Queen
    and me.
Then fill up the glasses as quick as you can,
And sprinkle the table with buttons and bran:
Put cats in the coffee, and mice in the tea—
And welcome Queen Alice with thirty-times-three!

            —Lewis Carroll

Chances are, your system can handle lowercase letters. Even if you do nothing else to improve your screen design, using upper- and lowercase alone makes the screens much easier to read.

- *Error message location.* Traditionally, error messages appear at the bottom of the screen. But consider the users' eye movement. They fill in the screen information and press enter. Then, expecting the next screen to appear, their eyes naturally move to the top left corner of the screen. When an error message displays, they must move their eyes back down again to the bottom of the screen. Many human factor specialists suggest putting error messages at the top to save eye movement. A minor savings, perhaps, but it all adds up.

- *Indents.* The more unique indentations you have in your screen design, the harder it is to use. Minimize the indents by lining up fields and field labels. Combine field labels of similar lengths, where possible, without sacrificing the logical order. And place the colons directly after the field labels.

This screen has multiple indents:

```
CUSTOMER INFO:
      NUMBER: 091851                         PHONE:   (215) 555-2600
      NAME: DAVID LIEBERMAN

ADDRESS:      STREET:   14 MAPLE TERRACE
CITY/STATE:   STRAWBERRY FIELDS, NJ      ZIP: 08099
```

The following screen, on the other hand, is easier to read. It has only two indents.

```
CUSTOMER INFORMATION
      NUMBER:    091851
      NAME:      DAVID LIEBERMAN
      PHONE:     (215) 555-2600

ADDRESS
      STREET:    14 MAPLE TERRACE
      CITY:      STRAWBERRY FIELDS
      STATE:     NJ
      ZIP:       08099
```

- *Highlighting.* Use highlighting for emphasis, but only to a point. Excess highlighting dilutes the effect. For example, overuse of blinking, color, and reverse video can make screens "noisy." Instead of getting the users' attention, you give them sensory overload. To prevent this, use lower levels of highlighting most of the time. Save the more powerful highlighting for when you really need to get their attention. Figure 6–12 lists some highlighting methods in order of intensity.

- *Headings.* Many screen headings are too complicated and contain superfluous information. They are filled with information that is there just because it's always been there. Certainly some of this information is necessary, but some of it may just be taking up valuable screen space. For example, do users really need to see the date, time, and internal program number at the top of every screen? You can often simplify these headings.

Designing a screen with human factors in mind saves a few seconds here and a few spaces there. All these nit-picking adjustments can add up to significant improvement over the lifetime of the screen. However, introducing these changes into your organization can be difficult. You will find resistance. People like things the way they are. Users can become

Most	blinking
intense	color
	size (e.g., large lettering)
	reverse video
	boxes
	underline
	boldface
	dim (lower intensity)
	uppercase
Least	brackets, parentheses, and asterisks
intense	white space (indents and blank lines)

Figure 6-12 Highlighting techniques in order of intensity

comfortable with meaningless field labels. And programmers find it easier to work in all uppercase mode. You may get resistance from your technical staff as well. They might tell you, for example, that lowercase lettering is impossible with the existing hardware, while further research would reveal otherwise. Try to overcome this resistance. People eventually adjust to the changes and productivity will increase.

Now that the screens are designed and documented, you are ready to design the reports.

REPORTS

This part of the agenda defines all output from the system. Besides standard reports, this includes any other printouts generated from the system, such as invoices, statements, checks, and labels.

REPORT LIST

Orders by Date:	Lists all orders (sorted by date) in the date range specified.
Orders by Customer:	Lists all orders (sorted by customer) in the date range specified.
Back Orders:	Lists all orders (sorted by product number) that have not been filled because the items are not in stock.
Sales by Region:	Lists all orders (sorted by region) for the current month and year to date.
Customer List:	Lists all customers and their profiles. These are leads that the salesman can follow up on.

Figure 6-13 Report list

Before the Session

Interviews before the session can generate:

- report descriptions
- samples of existing reports
- prototypes of new reports

The following describes how this information was put in the Working Document.

Report Descriptions

In their most abbreviated form, report descriptions list only report names and a brief summary of their functions, as shown in Figure 6–13. On the other hand, detailed report descriptions include more information. Figure 6–14 shows a detailed report description for one report.

REPORT DESCRIPTIONS

Report name:	Customer List
Description:	Lists all customers and their profiles.
Frequency:	Monthly
Copies:	3
Distribution:	Sales Manager Marketing Manager Order Entry Supervisor
Selection:	All customers that have made purchases within the last five years.
Sort:	Sort alphabetically by customer name.
Data elements:	Customer Number Customer Name (first and last) Customer Address (street, city, and zip) Customer Phone Number Region Code Discount Rate Credit Limit Terms

---

Figure 6-14 Detailed report description

Samples of Existing Reports

This includes samples of all the reports that the system currently generates. Pick one-page samples that are typical and that clearly show how the reports sort and which totals are included. For reports that end in summary pages, show the last page as well. When assembling these samples, do not overlook the obscure reports (such as the one used by the guy in the far corner of the lower level of the building).

Prototypes of New Reports

These are samples of how new reports might look. Usually, before the session, not enough specifications have been defined to build prototype reports. Since data elements and work flow have not been finalized, it may be better to wait until the session to do any actual report *design*. More than likely, you begin with only report *descriptions*, as shown above. If, however, users are clear about how the output should look, prototype reports can be designed and included in the Working Document. You can show just the column headings across the top or you can fill in the body with sample data or symbols that show field size. Examples of symbols are:

- xxxxxx—indicates text
- 999999—indicates numbers
- mm/dd/yy—indicates dates

The following shows a report design using these symbols.

MM/DD/YY	ORDER PROCESSING SYSTEM SALES BY REGION		PAGE 1
REGION 99			
ITEM NO.	DESCRIPTION	MTD ($)	YTD ($)
9999999	XXXXXXXXXXXX	999,999	9,999,999
9999999	XXXXXXXXXXXX	999,999	9,999,999
9999999	XXXXXXXXXXXX	999,999	9,999,999

For both existing and prototype reports, you have prepared overhead transparencies.

During the Session

If report descriptions have been prepared before the session, review the report names and definitions. If you are starting with nothing, ask the group, "What reports do you need?" On a flip chart, write the report names and short descriptions. The result is the report list shown in Figure 6–13.

After the group agrees on the reporting needs, complete the detailed report descriptions. Begin by deciding which of the following to include in the description:

- report name
- description
- frequency
- number of copies
- distribution list
- selection criteria
- sort criteria
- data elements

Then fill in this information for each report. The scribe notes the specifications on a Report Description scribe form that has spaces for all the items that can be included in a report description. The result is the same as shown in the report description sample in Figure 6–14.

Having described the reports, you are prepared to design the formats. You will either update existing reports or define new ones.

Updating Existing Reports

Review the reports from the existing system to see if they will be needed in the new system. Sometimes adding a column or changing some field lengths is all that is necessary. Say, for example, some users want an Orders by Date report modeled after the one used in the Denver division. They want the same format, but with an added column to show order status. Or they might want the existing report to stay as it is, but also to use it as a basis for a new report.

When the group has identified which existing reports they want to keep, give them one more chance to delete some. Here's where you trot out the speech about killing trees to make paper to print reports that collect dust on a shelf. Ask, "Do you really need all these reports? If you are using only part of one report, could that information be combined with another, more frequently used report? Could the report be micro-fiched?" There are laser printers that kick out more than 120 pages per minute. There are continuous form printers that generate 190 pages per minute. That translates to more than a quarter million pages per day! How the business world loves those printed reports!

Overhead transparencies work well for updating existing reports. Project the transparencies and guide the participants through the changes. If changes are simple, mark them directly on the transparency so people can clearly see what has been decided. If changes are more complex, project the transparency on the board and make the changes directly on the board. This might be easier than using wet paper towels to smudge away the pen marks from the transparencies. The scribe documents all changes on the existing reports within the Working Document.

Defining New Reports

If you prepared prototype reports before the session, review and update them using overheads in the same way as you handled existing reports. Defining new reports from scratch is similar to designing screens. Follow these steps:

1. *Set up the board.* Draw a large empty frame (representing the report) on the board. Arrange the magnetics of data elements (defined earlier) off to the side on a separate board panel.

2. *Define the headers.* (Reports generally do not have footers.) Write the header on the board at the top of the report frame. Add the report title.

3. *Define fields for the report.* Move the data element magnetics across the width of the report, indicating which fields print on the report. (This is the information that prints below the column headings.)

4. *Define the column headings.* You can do this in the same way that you defined field labels for the screens. For example, column headings might be Customer Name, Customer Number, Item Name, and Quantity.

5. *Fine-tune the placement of the fields and column headings.* Consider space limitations. Most reports are limited to 80 characters for narrow pages (8-1/2 by 11 inches) or 132 characters for wide pages (11 by 14 inches).

6. *Add summary text and totals.* In a report sorted by region, for example, you may want to include subtotals for each region as well as grand totals at the end. Summary text can be added to help clarify the subtotals. Again, you can use sample data or symbols (such as xxx, 999, or mm/dd/yy) to fill the body of the report.

The result is a completely designed report containing:

• the header (including the report title)

- column headings
- fields shown as either symbols (such as xxx) or data elements on magnetics indicating what is contained in each column
- summary text and totals

The scribe documents the completed report design on the Report Design scribe form. Figure 6–15 shows a completed sample.

OTHER AGENDA ITEMS

Although the following agenda items can be included in JAD sessions, they are not covered as consistently as the ones described so far. Usually, they are defined after the JAD. Furthermore, it seems that when they do arise, they are handled in a different way each time depending on what is needed for the system. They are summarized on the next page.

APRIL 20, 1990	ORDER PROCESSING SYSTEM SALES BY REGION		PAGE 1
REGION 30			
ITEM NO.	DESCRIPTION	MTD ($)	YTD ($)
9999999	XXXXXXXXXX	999,999	9,999,999
9999999	XXXXXXXXXX	999,999	9,999,999
9999999	XXXXXXXXXX	999,999	9,999,999
9999999	XXXXXXXXXX	999,999	9,999,999
9999999	XXXXXXXXXX	999,999	9,999,999
9999999	XXXXXXXXXX	999,999	9,999,999
9999999	XXXXXXXXXX	999,999	9,999,999
9999999	XXXXXXXXXX	999,999	9,999,999
9999999	XXXXXXXXXX	999,999	9,999,999
9999999	XXXXXXXXXX	999,999	9,999,999
9999999	XXXXXXXXXX	999,999	9,999,999
9999999	XXXXXXXXXX	999,999	9,999,999
9999999	XXXXXXXXXX	999,999	9,999,999
9999999	XXXXXXXXXX	999,999	9,999,999
9999999	XXXXXXXXXX	999,999	9,999,999
9999999	XXXXXXXXXX	999,999	9,999,999
TOTAL SALES FOR REGION 30		999,999,999	999,999,999
GRAND TOTALS		999,999,999	999,999,999

Figure 6-15 Completed report design

Records

This can include such items as record descriptions, record volumes, and logical design. See the segment called "One JAD, Another Hundred Data Elements" in the "Specialized Projects" section of Chapter 12.

Transactions

This defines transactions needed for the system, including a definition of the transaction and a list of all data elements used in the transaction. See the segment called "A JAD of Transactions" in the "Specialized Projects" section of Chapter 12.

Processing

This can include defining calculations and edits. See the "Detailed Edits in a JAD?" section of Chapter 12.

Manual Forms

This involves defining any new forms required by the new work flow. For example, for the Order Processing system, you might want to design a manual form for agents to:

- capture new customer information. The form would include lines for such information as customer name, billing address, shipping address, phone, and terms.
- capture change of address information for their customers.

The scribe documents the manual forms information on the Manual Form Description form (for form name and description) and the Manual Form Design form (for actual form design).

OPEN ISSUES

This is the part of the session where you address all those open questions that you have "put off until later."

Before the Session

Like assumptions, open issues have been accumulating since the start of the project. The Management Definition Guide includes all the issues that were identified up to that point. The Working Document contains those issues along with additional ones that arose after that. For example, in the following list, the first issue was part of the Management Definition Guide, while the last two are new ones that came up after that document was distributed. All three issues are included in the Working Document.

---

1. What are the requirements for bringing microfiche readers into the department?

2. To identify who took which orders, should we use the same clerk codes as we do today, or should we set up new ones?

3. Who will have access to the system? Will this access be defined at the screen level?

During the Session

Open issues are added throughout the session. When you see that a discussion is going on for a while and the answer to that issue is not immediately needed, recommend making it an open issue. For example, say the participants are discussing screens, particularly the data element called Customer Number. The conversation migrates to format. The discussion might go like this:

User 1:	We can put the customer number as the first field on the screen, followed by customer name and address.
MIS:	How long is the customer number field?
User 1:	Seven digits.
User 2:	But we use eight digits and the Boston division uses one letter followed by six digits.
User 1:	Well, it seems more people use the seven-digit format, so shouldn't that be the standard?
	This is where the leader steps in.
Leader:	Since we do not have enough people here to represent the various customer number formats, perhaps we could make it an open issue. At the end of this session, when we discuss open issues, we can identify who will meet to resolve this issue and when. Meanwhile, we can get back to the original question about what information displays at the top of the screen.

At this point, someone from the group puts the issue into words while the scribe documents it in detail on an Open Issue scribe form. The leader writes a one-line summary of the question on the Open Issues flip chart:

OPEN ISSUES

1. What will be the standard format for customer numbers?

Then the session moves on.

At the end of the session, all issues are reviewed. For each one:

1. The scribe reads the issue and the leader opens it up for discussion.

2. If participants can resolve it, the scribe documents the agreement on an Assumption scribe form.

3. If they cannot resolve it, the question is left as an open issue. Determine who will resolve it and when it should be resolved.

4. If more than one person is assigned to an issue, designate an issue coordinator.

For the customer number issue, Figure 6–16 shows how the completed scribe form might look.

OPEN ISSUE

Issue number:
Issue number: 1

Issue name: Standardizing customer numbers

Assigned to: Allison Brooke (coordinator)
Arthur Dent
Barbara Hurter
Louis Lestat
Hazel Rah
Jacquelynn Rudolph

Resolve by: 9/15/90

Description: What will be the standard format for customer numbers? The current formats in the six divisions are:

Boston: X999999

Denver: 99-999-999

Hoboken: XX999

Miami,
Milwaukee,
and Seattle: 999-9999

Once the numbers are standardized, all customers will have to be renumbered to prevent duplicate numbers.

Figure 6-16 Completed Open Issues scribe form

Sometime after the session, these smaller groups meet on their own to discuss the issues. The JAD leader is not involved. When an issue is resolved, the issue coordinator sends a written copy of the outcome to the executive sponsor and to all the people on the distribution list. Meanwhile, the executive sponsor monitors the resolution of these issues and follows up on those not resolved by the date determined in the session.

THE EVALUATION FORM

The evaluation form measures the success and participant satisfaction, not of the system design, but of the JAD session itself. This evaluation allows the person managing the JAD methodology to monitor how the users and MIS perceive the sessions, as well as how the leader and support person are performing.

The evaluation form is most beneficial when you first bring JAD into the company. You especially need this feedback for the first few projects.

As the use of JAD progresses and the methodology becomes an accepted tradition, you still need to keep this feedback mechanism in place. Although you may not ask the participants to fill out evaluation forms after every session (after all, sometimes the same people participate in several JADs), you might want periodic feedback at least twice a year.

When Should You Pass Out the Evaluation Form?

Most evaluation forms at seminars or conferences are distributed at the very end, when people are looking at their watches and thinking about lunch or dinner. If you pass out the evaluation form at this time, you will not get helpful feedback.

Instead, pass out the evaluation forms on the last day of the session, but before the last break. Ask the participants to take a few minutes to give you their comments on how JAD has worked for them. Figure 6–17 shows a sample evaluation form.

Then, after the session, read the evaluation forms carefully. Evaluate the comments and use what you can to rework and improve the next JAD session.

CLOSING THE SESSION

At the close of the session, you need to:

- determine who receives the final document
- discuss how the participants will review the document
- give some closing remarks

JAD EVALUATION FORM

Name (optional) _____ Date _____

JAD Project _____

1. Any comments for improving the Working Document?

    _____

    _____

2. During the session, which agenda items were not covered adequately?

    _____

    _____

3. Which agenda items were covered in too much detail?

    _____

    _____

4. How would you rate the visual aids (magnetics, overheads, and flip charts)?

    ____ Excellent   ____ Good   ____ Satisfactory   ____ Poor

    Why? _____

    _____

5. How would you rate the facilities used for the session?

    ____ Excellent   ____ Good   ____ Satisfactory   ____ Poor

    Why? _____

    _____

6. How would you rate the performance of the session leader?

    ____ Excellent   ____ Good   ____ Satisfactory   ____ Poor

    Why? _____

    _____

7. On the back of this form, please write your suggestions for improving any part of the session.

Figure 6-17 Evaluation form for the JAD session

Determine Who Receives the Final Document

With everyone together in the room, finalize the list of who receives the final JAD Design Document. Have them refer to the page in the Working Document that shows this list. Figure 6–18 shows a sample.

Explain that this list includes all participants and the executive sponsor. Ask if anyone else should be added to the list. If someone needs extra copies (for example, someone wants copies for two members of his or her staff), change the "No. of Copies" column as follows:

Name	No. of Copies	Mail Code
Linda Morgan	3	1

This way, the participant asking for extra copies can pass along the documents rather than you sending it directly to the new name. This simplifies the distribution process for you and prevents people from receiving a document without an explanation of what it is, where it came from, and why they received it.

After updating the names and confirming the mail codes, you have a finalized distribution list.

Distribution List for the Final Document

The final document will be sent as follows:

Name	No. of Copies	Mail Code
Peg Barry	1	2
Allison Brooke	1	3
Marko Chestnut	1	3
Martha Clancy	1	7
Abby Eron	1	3
Michael Kowalski	1	4
Linda Morgan	1	1
Charles Mugler	1	9
Ruth Noble	1	3
Anna Schwartz	1	4
Jean Willis	1	2
Sarah Wood	1	9

Figure 6-18 Distribution list for the final document

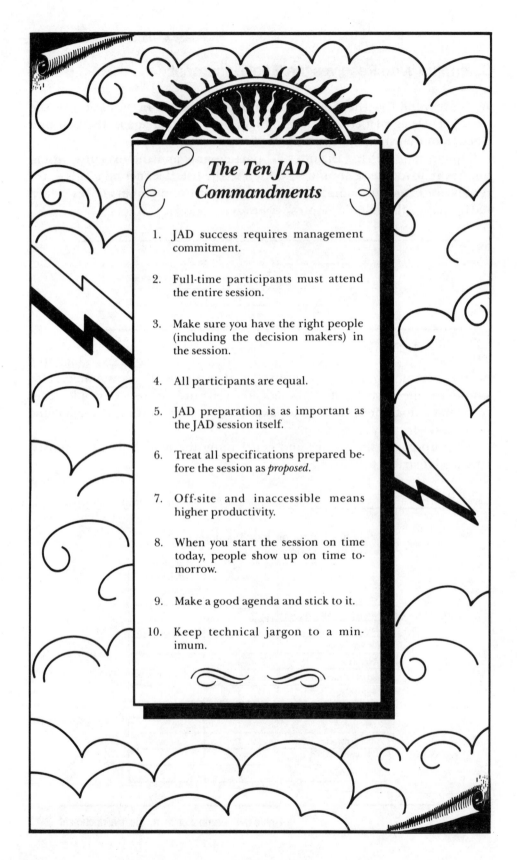

The Ten JAD Commandments

1. JAD success requires management commitment.

2. Full-time participants must attend the entire session.

3. Make sure you have the right people (including the decision makers) in the session.

4. All participants are equal.

5. JAD preparation is as important as the JAD session itself.

6. Treat all specifications prepared before the session as *proposed*.

7. Off-site and inaccessible means higher productivity.

8. When you start the session on time today, people show up on time tomorrow.

9. Make a good agenda and stick to it.

10. Keep technical jargon to a minimum.

Discuss the Review Process

Describe how the document will be reviewed and finalized. Tell the group that they will receive a copy of the document for their review. Then you will all meet again to discuss the changes people would like to make to the document. Once the changes are agreed upon, key participants sign an approval form.

At this point in the session, discuss who should be designated to sign the approval form. These signatures will represent consensus of all participants. Suggest one or two key user managers and the MIS project manager. These signatures will be obtained only after the review meeting, at which time everyone has agreed with the document contents. See the "Approving the Document" section in Chapter 7.

Closing Remarks

It is 12:25. Stomachs are empty and growling. Heads are filled with screen fields and work flows. The final moments are here. The agenda is complete. The session is over, almost. All that is left is for you to say a few parting words. You want to leave them on a positive note, thanking them for their efforts. You might say:

Okay, that wraps it up. I think this has been a really productive session, and I hope you're as satisfied with the results as I've been with your commitment and enthusiasm. The draft document will be out in about a week, and I look forward to seeing you at the review meeting. Thanks for your input.

THE TEN JAD COMMANDMENTS

A successful JAD requires good planning before the session, carrying out that plan during the session, and following through with a quality design document after the session. If the keys to success could be reduced to a few laws, they might read as does the list on page 130.

Having completed the JAD session, you are ready for Phase 5: The Final Document.

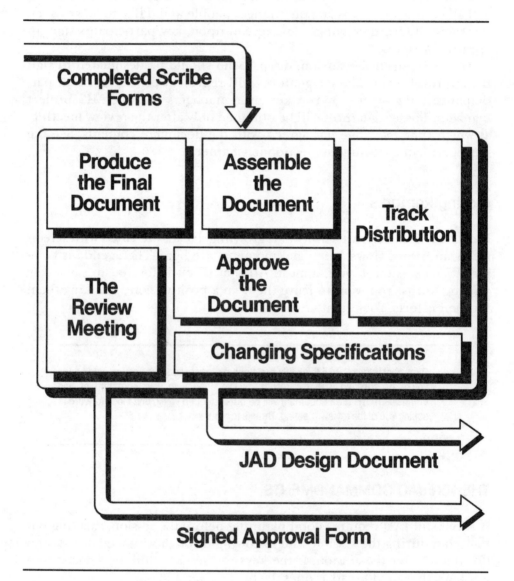

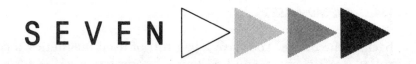
PHASE 5: THE FINAL DOCUMENT

When the session is over, you have all the information you need. You have gotten through the hardest part of the project and the part you probably had the most uncertainty about—getting all the participants to agree on one system design. Having done that, you naturally have a feeling of accomplishment. It is easy to say "Ah, we did it!" and simply let up on the project throttle. Don't do it. A critical part still remains. You must put all the information that came out of the JAD session into a format that can be used by the programmers, users, and all those involved in the next phase of system development.

In this, the final JAD phase, you transfer all the agreements made in the session into the final document. You assemble and distribute that document to the participants for review. Finally, you get signatures to approve the final document and authorize the development team to begin program design and coding.

PRODUCING THE FINAL DOCUMENT

The session is over, the boards are wiped clean, and the participants (filled to the brim with Danish and data element definitions) are headed back to the office. You have a bag full of used magnetics, a collection of rolled-up flip charts, and a stack of completed scribe forms. Now you need to turn all this material into the final JAD Design Document.

First of all, you can throw away the flip charts and wash the magnetics clean because all that information has been documented by the scribe. What you now have are the completed scribe forms and a Working Document filled with penciled additions and deletions. This is everything you need to produce the final document.

Why the Final Document Is So Important

The JAD Design Document is the culmination of all that has gone into the JAD project. It is a comprehensive synthesis of agreements made in the session. It is the one resulting document, the one final product, that represents JAD's role in the systems development process. For the people (particularly in upper management) who were not participants but have a line of responsibility for that project, the final document may be the only evidence they have to judge the status of the project after the JAD.

A good final document is the translation of user needs into system specifications. Unless these specifications are clearly and completely documented, they are lost as soon as the session is over. No one will remember the details of what the participants decided even two days ago. And after a month, you might even wonder, "What JAD session?" This highlights the importance of completing the final document as soon as possible after the JAD session is over. Presumably your scribe has taken good notes, but they are nevertheless notes and not complete texts. Your work will be harder if you go on vacation then return to the task. So get the document written. *Then* fly off to that island.

Producing a quality final document requires certain techniques in compiling the assumptions, work flow, screens, reports, open issues, and whatever else comes out of the session. The following describes how to do this.

Converting the Working Document

Since the Working Document was set up to mirror the final version, you already have a good starting point. You need only update that document with the additions and deletions noted by the scribe in his or her copy of the document. More than likely, you have the Working Document stored in files on a PC word processor or a mainframe text editor. To convert these files from the Working Document to the final version, change the following:

- title page
- preface
- section introductions, if necessary

The following describes these changes.

Title Page

Make these changes to the title page:

1. Replace "Working Document" with "JAD Design." (Leave the system name as it is.)
2. Replace the Working Document date with the final document date.
3. Add the names of the participants in alphabetical order.

JAD DESIGN
ORDER PROCESSING SYSTEM

July 17, 1990

Allison Brooke
Marko Chestnut
Martha Clancy
Abby Eron
Michael Kowalski
Linda Morgan
Charles Mugler
Anna Schwartz
Jean Willis
Sarah Wood

Preface

Replace the Working Document preface. The new one might say:

PREFACE

This design document describes user requirements for the Order Processing system. It includes all specifications defined in the JAD session held July 7 to 10, 1990. When the participants approve this document, MIS can continue program design based on its contents and the resolutions of open issues.

Section Introductions

Since you prepared the Working Document before the session, there will be references to the session in the future tense. This means you need to change any comments about what *will be* done in the session to what *was* done in the session. For example, in the Agenda part, change "The following agenda *will be* accomplished in the four-day JAD session" to ". . . *was* accomplished . . ."

Organizing the Source Documents

To organize the source documents, work at a large table. Sort the scribe forms into sections and subsections. Consider such things as: Do you want the data elements in alphabetical order? Should the screens be grouped by function, or do you want all the existing screens first, then the new ones? How will you present the reports?

By organizing all the information into the order you want, you will have determined the complete table of contents. (The page numbers will be filled in later.) It might look like this:

TABLE OF CONTENTS

JAD overview xx

Agenda xx

Session agenda xx

Session participants xx

Distribution of the final document. xx

Assumptions xx

Now you are ready to expand the table of contents into the final document.

Entering the Information

With the scribe forms organized and the table of contents complete, you can enter all the information from the scribe forms into the document. Standard template files can be set up for this step. In other words, you can build templates for various parts of the document. Then you copy these templates and add the variable information from the scribe forms. Chapter 10 describes how to do this.

Enter the text carefully, making sure it is correct and complete. But, just in case you missed anything, the next step can take care of it.

Editing

After entering all the information from the scribe forms, you have the first draft version of the final document. This version goes through edits to assure that it is clear and consistent with the scribe forms. This involves two separate edits:

- editing for clarity
- editing for accuracy

Editing for Clarity (the Grammarian Slaughter)

With a critical eye, review the document from the perspective of the reader. Who is your audience? Will the document be understood by the users (who are more procedure-oriented) *and* by MIS (who is more systems-oriented)? Is the text written in clear English? (Or is it loaded with unexplained technical jargon?) Are acronyms spelled out at their first mention? (Or do you say things like "The INV records of FAS must be revised to accept the MUN-SPLIV files.") If you have created tables in the document, are those tables titled and labeled with meaningful column headings? These are the details that are sometimes overlooked in the scribing process because everything seems so clear at the time.

Editing for Accuracy (the Compulsive Edit)

This is the edit where we could be accused of going too far. Neverthe-less, every time we do it, we find an error or two that makes it worthwhile.

This edit assures that everything on the scribe forms has gone into the final document and that it is accurate. Close yourself in a quiet room. (The job is tedious, and if you are distracted, you will miss the errors.) With the sorted scribe forms on one side and the document on the other, check each form against the document. In the data elements section, for example, verify that each data element is there *and* check that such details as length and description have been entered correctly.

With all the changes made from these edits, you have completed a draft copy of the final document. Now it must be distributed and reviewed by the participants.

ASSEMBLING THE FINAL DOCUMENT

To prepare the document for distribution, make enough copies for all the participants. At this point, copies are not sent to the additional people who requested them, just to those who will review the document (in other words, the session participants).

Put the copies into three-ring binders. You can use one-inch binders with clear plastic for holding preprinted covers. This allows you to insert front and back covers as well as a spine. These inserts can be set up as follows:

- *front cover*—the main title (JAD Design) and company name (for example, Gadgets Galore, Inc.)

Figure 7-1 Binder for the final document

- *back cover*—company name and logo
- *spine tab*—project name (Order Processing System) and title (JAD Design)

Figure 7–1 shows how the binder would look.

The front and back covers are generic and can be used for any JAD projects. The spine tab is customized by including the project name, such as Order Processing System. Without this customized spine tab, some participants (who have been involved in several JADs) will have a row of binders that cannot be distinguished.

Within the document, you can separate the sections with preprinted tabs. Figure 7–2 shows how these tabs might look.

The full package includes the binder (with covers and spine tabs), pages, tabs, and a cover memo. A sample cover memo is shown in Figure 7–3.

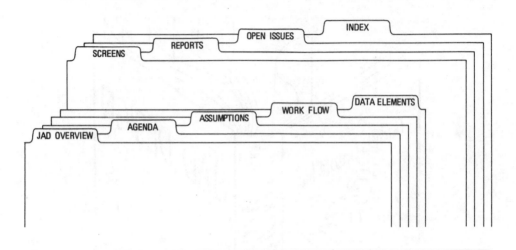

Figure 7-2 Tabs for the final document

TRACKING DISTRIBUTION

As you can see, a JAD requires several mailings to several groups of people. You need a way to track this distribution so that:

- you can easily delegate distribution tasks without having to give complex explanations (for example, you might have a secretary assemble and distribute the documents)

July 17, 1990

To: Order Processing JAD Participants

From: Martha Clancy

Subject: JAD Review Meeting

Enclosed is a draft copy of the final JAD Design Document for your review.

To assemble your document, please insert the tabs before each document section. (For example, insert the *Assumptions* tab before the Assumptions section.)

A post-design meeting has been scheduled to discuss changes and to approve the document. The meeting will take place Monday, July 20, at 10:00 in the main conference room.

It is important that all participants attend. If you have a conflict, please let us know. Otherwise, see you Monday.

Figure 7-3 Cover memo for document review

- you have a history of who received what
- you can determine inventory requirements for binders, cover in-serts, and tabs

Figure 7–4 shows a form that allows you to track distribution.

Different JADs require tracking different information. Record only what you need. There is no sense tracking numbers for statistics' sake alone.

THE REVIEW MEETING

This one- to two-hour review meeting is held in an on-site conference room. The JAD leader presides and all participants attend. Even those who are in complete agreement with the document as is should be there, because they could be affected by other people's changes.

In the meeting, highlight the fact that although specifications will continue to change, this document represents a specific point in time, that point being the close of the JAD session. Specifications that have changed since then (such as an open issue that has been resolved) are not reflected in this document. If the users want to add a new field to a report, for example, they would discuss it with MIS. Making this clear

Document Distribution Form

Name	Mail Code	Mgt Def Guide	Working Doc	Draft Doc	Final Doc
Peg Barry	2		X		
Allison Brooke	3	X	X	X	X
Marko Chestnut	3	X	X	X	X
Martha Clancy	7	X	X	X	X
Abby Eron	3	X	X	X	X
Spencer Hoag	9	X			X
Michael Kowalski	4	X	X	X	X
Linda Morgan	1	X	X	X	X
Charles Mugler	9	X	X	X	X
Ruth Noble	8	X	X		X
Anna Schwartz	4	X	X	X	X
Marilyn Segal	4	X			X
Jean Willis	2	X	X	X	X
Sarah Wood	9	X	X	X	X

Figure 7-4 Document distribution form

prevents new discussions that could turn a simple review meeting into a mini-JAD forming in the wake of the main session. After all, these same participants just spent days, closed in a room, in an active decision-making mode. They can easily fall back into these roles. It is your job to keep them on track.

Walk through the document page by page. You will hear a variety of comments that involve various kinds of edits. The following gives examples of these different types of edits:

- *Accuracy edits.* "This is not right. It should say . . ."
- *Clarity edits.* "Assumption number 3 is confusing. Can we change the wording to . . . ?"
- *Micro edits.* "Wouldn't it be better to say '*the* data base' instead of '*a* data base?'" (There is often someone present who is interested in such details.)
- *After-the-session thoughts.* "I was thinking we should add another column to the Customer List report . . ."

For the accuracy and clarity edits, get agreement from the group and note the changes. These are the kinds of comments you want from the participants. For the micro edits, go ahead and make the changes, but do not take time discussing them. For the after-the-session thoughts, explain again that the document represents agreement at the time of the session. Suggest handling the changes through MIS.

As participants make their comments, restate them as you make your notes in the document. Ask the group if they agree. Then move on.

At the end of the meeting, determine how to handle the changes. In other words, should the document be reissued? If the changes are minimal (for example, the wording has just been fine-tuned on several pages), then the current version (plus their noted changes) can serve as the final copy. On the other hand, if the changes are significant (several changes have been noted that will affect the programmer's coding), you need to update the document and send it out again, along with the cover memo shown in Figure 7–5.

APPROVING THE DOCUMENT—THE FINAL OKAY

Approval represents total agreement with the contents of the final document. This includes the changes discussed in the review meeting. The people who will sign the approval form were designated in the session. If the changes in the review meeting are minor, then the end of that meeting is a good time to get these signatures because:

July 24, 1990

To: JAD Session Participants

From: Martha Clancy

Subject: JAD Design Document

   Enclosed is the final copy of the document. It includes updates from the review meeting.
   Please remove your current version and replace it with this copy.

Figure 7-5 Cover memo for the final document

- the people who will be signing are present at the meeting. This prevents having to rely on the uncertainty of interoffice mail, especially when several people are on the route.
- as specifications change, people may be less likely to sign, even though the document represents a point in time before these new specifications arose.

JAD APPROVAL FORM

JAD Project: Order Processing system

Authorizing User Managers:

_____ Date _____

_____ Date _____

_____ Date _____

Authorizing MIS Managers:

_____ Date _____

_____ Date _____

The above signatures represent joint agreement by the JAD participants as to the specifications for the Order Processing system. These include the contents of the JAD Design Document and the changes discussed in the review meeting. With this approval, MIS can continue program design based on this document.

Figure 7-6 Approval form

If the changes from the review meeting are major, however, wait until after the final version is distributed. Then, send the approval form to those designated. The signed approval form resides in the document received by the manager of Applications Development. Figure 7–6 shows a sample approval form.

With the review complete and the approval form signed, the JAD is done. Now you are ready to begin your next JAD project. Before that, however, perhaps you can arrange a well-deserved, one-week vacation on the Riviera.

CHANGING SPECIFICATIONS AFTER THE JAD

When the programming department takes over defining program specifications and coding, the JAD leader and staff no longer maintain the document. Nevertheless, specifications continue to change. The document represents a point in time. MIS needs a way to update specifications as new user requirements arise.

Sometimes, the people in programming can use the JAD Design Document as a base to update specifications as they change. In this case, copy the document files over to their area. The programmers will maintain them. The original files remain with you.

Keep your files for at least six months, since you may want to use parts in future documents. And always keep a hardcopy original (or a backup tape or diskette), because just when you delete the files from the computer, it will be your luck to have a team of newly hired consultants arrive and request five copies.

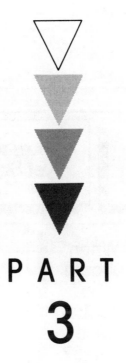

PART
3

MORE ABOUT JADS

Getting the Right People

How to Be Flexible yet Stay on Course

When Should the JAD Leader Interrupt?

When Should the Scribe Take Notes?

Minimizing Technical Jargon

How to Handle Indecision

How to Handle Conflict

Encouraging Shy Users

Chilling the Dominator

Stifling Sidebar Conversations

Humor in JADs

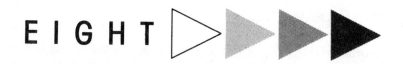

EIGHT

JAD PSYCHOLOGY

This chapter deals with the psychology involved in pulling off a successful session. It also discusses some of the problems in group dynamics that can seriously threaten success, and tells you how to deal with them. And it describes some ways to prevent these problems in the first place.

HAVING THE RIGHT PEOPLE IN THE ROOM

Consider the consequences of not having the right people at the session; that is, you do not have participants with the authority and knowledge to determine specifications for the system. This defeats the whole concept of JAD, which is to provide a concentrated workshop for making decisions with everyone present to make those decisions. If all the required people cannot attend, consider revising the JAD schedule.

To assure you do have the right people, spend time in the early phases selecting the participants. Ask the users who should attend. Ask the executive sponsor as well as the MIS director. When you meet with various people during the interview process, review the list of participants to see if they feel anyone has been overlooked.

On the other hand, you don't want to overload the session by adding another name every time one is mentioned. It is all too easy to fall into the trap of "inviting" people for essentially political reasons, so they don't feel "left out." The JAD session is not a party. Investigate to make sure of the need for each person's attendance. Remember, some people can be designated as "on-call" to be contacted when questions arise in their area of expertise.

But What If You Missed Someone?

What do you do if you have overlooked someone who should have been included in the session? For example, several times in the session, people say, "We can't make *that* decision. Betty handles that area so it is up to her." Well then, it's time to call Betty and get her into the session. If you have selected the correct executive sponsor, he or she will support you in adjusting any priorities that Betty might have. If she cannot attend, make one open issue that lists all the unresolved questions where her input is essential. At the very least, talk with her on the phone to address the issues. Then, bring the resolutions to the next day's session.

HOW TO BE FLEXIBLE YET STAY ON COURSE

This sounds like a contradiction—being flexible and staying on course. But the two are not mutually exclusive. You can do both, but to keep the balance between remaining flexible and maintaining your course, you must be able to distinguish between unproductive digressions and necessary discussions.

The natural progression is this: The group is discussing an item on the agenda. While elaborating on a particular point, someone brings up a related subject. Someone else picks up on it. The conversation digresses. The original subject is lost. You have now gone from discussion to digression to being totally off the course.

The problem is, digressions are often laced with much fervor and zeal. An apparent urgency fills the air. This intensity sometimes clouds the leader's ability to recognize the tangents.

Throughout the session, you will sense such strayings from the agenda. Ask yourself, "Is the discussion necessary to accomplish this part of the agenda?" If you do not know, ask the group the same question. Sometimes they will say, "Yes, we need to cover this." In this case, continue the discussion. Other times, they will cease their meanderings and get back on track. Most people in the room, having tolerated the digression in silence, will be glad to get back to the questions at hand.

The agenda is like a road map. You follow it through the course of the session. In a typical road trip, sometimes you make intentional side trips (the car breaks down and you need a new carburetor). But you still have the same destination. Other times, you get off course (you are lost) and migrate toward another destination (you head for the Grand Canyon instead of San Francisco). Now you have lost the original objective. The important thing in a JAD session is to recognize these migrations and get back on track.

Hidden-Agenda Digressions

Sometimes, people come to the session with pet gripes that have nothing to do with the session objectives. For example, one participant has, for two years, been trying to convince the home office to revise the way customer service handles complaints. This issue is not in the scope of the session. But as soon as the participant sees an opportunity, he tries to shift the agenda. Watch out for this. Bring the discussion back to the agenda at hand.

Merry-Go-Round Digressions

Sometimes, the group unanimously adopts a digression, professing that it is necessary. But the discussion continues in a circular fashion, leading nowhere. This is the time to make an open issue. Summarize the question. Document it on an Open Issue scribe form and write a one-line summary of the question on the Open Issues flip chart. Then come back to it later. If you cannot continue with that agenda item until the issue is resolved, move on to the next one. This is how you stay *flexible*, that is, being able to make necessary adjustments. The important thing is recognizing tangents and interrupting when necessary.

WHEN SHOULD THE JAD LEADER INTERRUPT?

The most obvious time for the leader to interrupt is when discussions go off course. Just as important as recognizing digressions, though, is recognizing *potential consensus*; that is, detecting when a design decision can be reached. This is what marketing people refer to as *closing the sale*. When a customer says, "I really think this car is right for me," the salesman does not ask, "Would you like to see the trunk?" Instead, he turns his focus to writing up the sale. You should use this same tactic at the JAD session. Suppose someone suggests, "Well, maybe we could convert all the various customer numbers to the format used by the Milwaukee division." And someone else says, "That could work." This is when you step in and say, "The suggestion is to convert the customer numbers to the Milwaukee format. Will this work for everyone?" If so, have the scribe document the decision and read it back to the group for consensus.

Another reason for interrupting is to call breaks. Sometimes the session stays right on course, decisions get made, and you have everyone's undivided attention. Nevertheless, the participants still need a recess. Try breaking on a positive note, such as after a decision has been made and documented. But do not drag the group through 20 minutes of restless, fidgety discussion just to get a decision. As they say, "The mind can only absorb what the seat can endure." And the seat can usually endure no more than two hours at a time. Football games have their halftimes. Baseball games have the seventh-inning stretch. And JAD sessions have their intermissions. Maybe *you* could go on for hours and hours. After all, you are up for the session. You are invigorated, doing your song and dance, writing things on the board, and generally operating at on-stage intensity. The participants, on the other hand, have been sitting in the same position all morning, drinking lots of coffee and brainstorming. So, do not lose track of the time. Interrupt to call the breaks.

On Time Today Means On Time Tomorrow

Now is the time for a word about when to start and end the sessions each day. The first day is a test. Most everyone will arrive on time, some will be a little late, and everyone will check to see if you really start at 8:30 A.M. or, as in many meetings, if you lollygag around for 20 minutes waiting for the late ones to arrive. How late or early you start on that very first day is their gauge for when to come the next day. The point is, *always* start on time! That way, you give the punctual ones the respect they deserve for their promptness. And you give the ones who come late the message they deserve, that you will not hold up the session for their fashionably late arrival. All participants are equal.

With this approach, participants will show up on time. Give them the same courtesy at the end of each meeting by adjourning on time. Do not hold them for even 10 more minutes. Close on time or a few minutes early. They will be ready.

WHEN SHOULD THE SCRIBE TAKE NOTES?

Very simply, the scribe takes notes when he or she receives a prompt from the leader. The scribe of a JAD session is not like the student of Sociology 101 who writes everything down because he or she doesn't know what will be on the test. Instead, the scribe can save pencil lead until you indicate what to write. Then, the scribe records either exactly what the participant says, or the leader's paraphrased version of it.

This is where you may need to slow down the person doing the talking. For example, when you ask the participant most familiar with the agreement to summarize it for the scribe, the participant might begin with an elaborate exposition on the subject and then jokingly ask, "Did you get all that?" Tell the participant to go slowly, as the scribe needs to record it. Make sure the scribe is bold enough to say, when necessary, "Could you please repeat that a little more slowly?"

KEEPING TECHNICAL JARGON TO A MINIMUM

The easiest way to confuse and alienate your users is to allow the MIS people to conduct a conversation amongst themselves laden with technical jargon, acronyms, and other gobbledygook. All you have to do is give the floor over to a pretentious hotshot data base tekkie rambling on, "In order for the application to perform properly, we need to operate the DBMS in an MVS/XA or ESA dataspace capability multi-processor JES2 shared-spool environment supported by BDAM or VSAM/RRDS access methods. . . ." The users will retreat into silence, totally annoyed.

Make sure everyone understands the language being used. Don't take for granted that users automatically understand all those catchy data processing phrases, such as *online realtime update, menu-driven systems*, and *navigating the data base*. If the technical people absolutely must use this jargon, have someone translate what they are saying into English so the rest of the group is not left in the dark. You, as the JAD leader, probably come from an MIS background and understand much of the technical jargon. You should therefore make a conscious effort to listen with the users' ears and to interrupt when you sense the MIS people are starting to talk over the users' heads.

MIS tends to look at a system in terms of inputs, outputs, files, and how data is updated. Users, on the other hand, are concerned with business activities. Since the system is being designed from the user's perspective, you need to speak more in terms of the business (insurance, manufacturing, and so on) and less in terms of the bits and bytes.

HOW TO HANDLE CONFLICT

Conflict comes in many forms. There is *gainful* conflict where a few people in the group explore and defend different ways to define part of the system design. This is productive and need not be curbed. There is *stalemate* conflict where the discussion comes to an impasse. You can go no further, maybe not even to the next agenda item. This must be dealt with. Finally, there is *dogmatic* conflict which spawns the willful, headstrong kinds of discussions where egos run rampant, blood pressures rise, and people take the attitude, "it's my way or the highway." This kind of conflict is unproductive and unnecessary. But when it arises, you have to squelch it immediately or it will wreck your session.

One of the jobs of the executive sponsor is to resolve conflicts when the session comes to a standstill. But you can and should try other approaches before things get to this point. The more the executive sponsor has to arbitrate, the more the results of the JAD session begin to take on the character of a top-down decision-making process imposed from above. We will describe several conflict-resolving techniques available to you. We suggest applying them in the following sequence. If one technique does not work, move on to the next.

1. Open Up the Question to the Group

Sometimes the conflict stems from outright disagreement. For example, one person feels that the format for customer numbers should remain unique to each division, while another feels they should be standardized. When you see no resolution in sight, change the perspective. Open up the question to the others in the group. Ask, "How do the rest of you feel about this? How about you, George, what would work best in your area?"

2. Make an Open Issue

The most common way to handle conflict is to make the question an open issue. This sets the conflict aside so you can get on with the agenda. By documenting the issue and putting it on the flip chart, the group feels confident that it will not be abandoned. Sometimes, in the process of documenting the issue, the very act of trying to put the question into words somehow brings an objectivity that leads to a resolution. More than likely, though, the issue stays as an open one.

However, when you need the resolution right then and there, talking about it tomorrow will not do, especially when everything else you need to cover today hinges upon or flows from the resolution of that particular issue. In that case, go on to the next technique.

3. Take a Break

When you are locked in a stalemate and the discussion appears unproductive, take a coffee break. During the interlude, people may get together in smaller group discussions. This more informal mode brings new perspectives. We have seen many heated conflicts resolved in the cooler environment that a break provides.

4. Analyze the Conflict in a Structured Way

Sometimes applying objective analysis can clarify the issues involved in a disagreement, and thus bring participants to consensus. To handle deadlocks, we use a technique that quantifies the impact of each side of the conflict. This six-step approach is described in the section called "How to Handle Indecision" on page 154. Meanwhile, if that does not work, proceed to the next technique.

5. Call the Executive Sponsor

When you cannot resolve the conflict within the confines of the session, it's time to call the executive sponsor. Usually, all the user departments represented in the room ultimately report to that person, so he or she has the authority to make decisions for their areas.

Explain the situation to the executive sponsor. If necessary, have someone from the group elaborate. Then, communicate his or her decision back to the group. The executive sponsor will probably resolve the issue very decisively: "The format for customer numbers will be . . ."

Conflicts between Users and MIS

Conflicts between users and MIS should be handled differently, because in this case the focus is on *who* is in the conflict rather than *what* the conflict is about. Sometimes you are dealing with relationships between users and MIS that are not exactly warm and cozy. For example, a programmer might carry on stubbornly about how a user's request absolutely cannot be done. Or a user might start badgering MIS about a particular point and bring up other frustrations like, "What do you mean you can't do it? And what about all those other outstanding requests that never get done?"

This is where you shift from analyst to psychologist. You need to make the programmer realize that the MIS department supports the users; therefore, if the users really need a certain function, then MIS needs to take a good look at how to do it. In the same way, you need to make the users understand the limitations of computer systems, and what may be realistically involved in implementing the requests they are asking for. In other words, try to bring both user and MIS perspectives to light and break down the barrier between them, even though at times this barrier may seem like the Berlin Wall.

HOW TO HANDLE INDECISION

Indecision is when the group just can't determine which way to go. In this situation, it helps to analyze the two or more alternatives you are trying to decide between.

For years, we have been using the decision-making technique we are about to describe. Recently, we learned it is similar to a technique that has an official name. Dr. Charles H. Kepner and Dr. Benjamin B. Tregoe, consultants in strategic and operational decision making, invented it and appropriately called it the *Kepner-Tregoe* approach. The modified form we describe here has worked well for us in JAD sessions and in other business situations as well.

Let's look at a quandary that most everyone has been in—deciding which automobile to buy. Let's say you have narrowed your search down to two cars, but cannot decide between the two. You like the Volvo and the Toyota. But which should you buy? Here is how to use the modified Kepner-Tregoe method to make your decision:

1. *List the alternatives.* In column headings across the top of the board, list your choices. These are the columns where you will be scoring each choice. The headings would be "Score (Volvo)" and "Score (Toyota)," like this:

Score	*Score*
(Volvo)	*(Toyota)*

If you were dealing with more than two choices, they would continue in separate column headings across the top, like this:

Score (Volvo)	Score (Toyota)	Score (Chevy)	Score (Chrysler)

2. *List the criteria.* Determine what is important to you in your selection. What do you want from an automobile? Your criteria might be:

- safety
- price
- gas mileage
- trunk space
- reliability
- color

List these criteria down the left side of the board:

Criteria	Score (Volvo)	Score (Toyota)
safety		
price		
gas mileage		
trunk space		
reliability		
color		

3. *Weigh the criteria.* Add a new column labeled "Weight Factor." Give each criterion a weight from 1 to 5 (where 5 is the best). For example, if you have two small children, *safety* might be the most important to you. Give it a weight of 5. *Gas mileage* may not be as significant because you will not be using this car for long-distance gas-guzzling trips. Give it a 3. *Color* might be the least important. Give it a 1. Fill in the others in the same way:

Criteria	Score (Volvo)	Score (Toyota)	Weight Factor
safety			5
price			3
gas mileage			3
trunk space			2
reliability			5
color			1

4. *Evaluate the criteria for each alternative.* At this point, you have read consumer reports and have been to the car dealerships. You have taken the test drive, asked all the standard car-buying questions, and heard how the manager is going to give you the best deal in town. Now you are ready to evaluate how each car meets the criteria. As you do this, assign a score of 1 to 5 (again 5 is the best). Write that number in the appropriate space in the grid.

   Beginning with *safety*, you have determined through research that the Volvo has an excellent safety record. Give it a score of 5. Although the Toyota may not be at the very top, it does have a good safety record. Give it a 4. Moving on to *price*, the Toyota, being less expensive, gains the better score. Continue evaluating all criteria. When the columns are filled in, the chart looks like this:

Criteria	Score (Volvo)	Score (Toyota)	Weight Factor
safety	5	4	5
price	3	5	3
gas mileage	3	5	3
trunk space	5	4	2
reliability	3	5	5
color	4	5	1

5. *Determine the weighted score.* By multiplying the weight factor by the score, you can determine the weighted score for each criteria. For example, to calculate *safety* for the Volvo, do the following:

5 (score)
× 5 (weight factor)
25 (weighted score)

For Toyota safety, the calculation is:

4 (score)
× 5 (weight factor)
20 (weighted score)

Continue calculating weighted scores for all the criteria. Make two new columns on the right side of the chart labeled "Weighted Score (Volvo)" and "Weighted Score (Toyota)." Fill in these two columns with your calculations. The chart looks like this:

Criteria	Score (Volvo)	Score (Toyota)	Weight Factor	Weighted Score (Volvo)	Weighted Score (Toyota)
safety	5	4	5	25	20
price	3	5	3	9	15
gas mileage	3	5	3	9	15
trunk space	5	4	2	10	8
reliability	3	5	5	15	25
color	4	5	1	4	5
Total Weighted Score:				72	88

6. *Total the scores.* Now you have only to total the scores in the "Weighted Score" column for each car. The highest score is the best choice based on the criteria you have set. Now you know which car to drive off the showroom floor, in this case, the Toyota.

This method can be effectively used to decide technical questions that arise in the session. For example, should we use unique customer number formats for each division or should we standardize them? In this case, the choices are *unique customer numbers* and *standard customer numbers*. The criteria might include:

• cost of standardizing formats
• ease of use for sales agents
• impact on system processing

To make this decision, guide the group through the decision-making process. As before, make a grid with choices and criteria. It might look like this:

Criteria	Score (Unique Customer Numbers)	Score (Standard Customer Numbers)
cost of standardizing formats		
ease of use for sales agents		
impact on system processing		

Now determine the weight factors, the criteria scores, the weighted scores, and the total scores which indicate the best choice.

Numbers Can Give a Sense of False Objectivity

Remember that the results of this kind of analysis are not necessarily the "last word." This method uses numbers in an attempt to give rational weight to what might otherwise be subjective value judgments. If the outcome conflicts sharply with intuitive judgments, it could be you have not given weights to the criteria that truly reflect the group's opinions.

If your calculations indicate one choice, but the group's intuition leans toward another, listen to that intuition. You might end up throwing out the scores and buying the Volvo because you like the image it projects. This could mean that you overlooked certain criteria in the analysis. Maybe you should have added a criterion called "image factor."

CHILLING THE DOMINATOR

If you hold many JADs, at some point you are going to run into the *Dominator*. You know, the one with the loud plaid jacket, the gritty voice, and the muscular brain. The Dominator's the one who has to speak first and get in the last word as well. He or she (they come in both genders) tries to take over the session. The Dominator says things like, "You can't do that, it won't work!" or "I have a better idea." And he or she speaks in paragraphs that have no end. Meanwhile, the rest of the group starts to daydream, doodle, and roll their eyes in annoyance.

The only way to deal with the Dominator is straight on. Explain to him or her that other points of view must be heard. When the Dominator has carried on long enough, say, "Now, how do the rest of you feel about this subject?" If the person persists in dragging down the session, speak with him or her at the break, explaining that this dominance is hindering the group effort. You might say, "Dan, we are fortunate to have your exper-tise. Your view is important to the session. But we also need to hear from others as well. Would you please give the other people more of a chance to participate? If we don't balance the session with all points of view, the resulting system design will not work."

If the Dominator persists, the choice comes down to this: Do you want to take the consequences of running a session that meets the needs of only one aggressive participant, or take the consequences of asking that person to leave the session? Depending on the political situation, you are usually better off having the person leave, unless you find yourself in the circum-stances discussed in the following section.

What If the Dominator Is Your Boss?

Now here is a problem for you: What if your boss is the one slowing down the session because of his or her dominant behavior and continuous digressions? Do you tell your boss to get back on track or "exit stage right"? Euphemistically speaking, this action could have a negative effect on your career.

If your boss is a dominating digressor, you probably know that already. Take the preventive approach. Do whatever you can to keep this person off the participation list. You can explain to your boss that his or her contribution would be most valuable before the session, in reviewing the preliminary specifications and contributing ideas at that time.

ENCOURAGING SHY USERS

The opposite of the Dominator is the *Shy User*. This is the person who probably comes to the session fully prepared, with several sharpened pencils and a tablet of paper. He or she knows the subject matter, has good ideas, but is too shy to say anything in front of all those people.

This is a problem you need to anticipate and plan for by learning ahead of time who has the expertise for each part of the agenda being discussed. For example, Marvin has a complete understanding of how orders are sorted and entered into the system. And Claire knows the technical side of how the existing system interfaces with other systems in the company.

You may come to a point in the discussion where you realize that Marvin has an in-depth knowledge of the question at hand, but he is not speaking. More than likely, his silence is not from being without opinions, but from being too shy to speak in this forum. Nevertheless, you need his view.

Draw him out. Ask him a direct question. Stay away from open general questions like, "Marvin, what do you think about this?" Instead, begin with a specific inquiry like, "Marvin, what process is used to sort orders before entering them?" Furthermore, ask about the department's view rather than his own personal view. For example, do not ask, "What do *you* do?" Instead, ask, "What does the Order Entry department do?" You must try to elicit information through effective questioning without pressuring or embarrassing the participant.

STIFLING SIDEBAR CONVERSATIONS

This particular problem relates to the concept of *group integrity*. In the session, the group must remain cohesive. The integrity of the group dissolves when subgroups emerge. A typical scenario is this: You are covering a controversial subject that affects many people in the room. One person has the floor, talking about a particular approach. Two others start a little conversation on one side of the room. Three others start talking on the other side. You have lost group integrity.

Maintaining this integrity means that only one person speaks at a time and he or she speaks to the whole group, not just the people sitting nearby. This way, no one misses anything. Keeping a room full of people focused on one conversation at a time for three to five days is a challenge. You need to watch for sidebar conversations and handle them as they arise.

When conversations do emerge, take the following two-step approach:

1. As you are leading the session, walk toward the people who are talking. As your voice approaches them, they will probably turn their attention back to the session. This physical approach works more often than not.

2. If the conversation continues, deal with it directly, but diplomatically. There is no need to admonish them for "talking in class." This would only make them resentful. Instead, you might say something like, "Renee. David. We need your input on this." Or, for something stronger, you could say, "Hey folks, can we get back on

track here, please?" Then, immediately turn your attention back to the session. You need only their silence, not their chagrin.

Address these sidebar conversations as they come up. You want everyone to hear what everyone else has to say. And above all, you want to keep the group together.

HUMOR IN JADS

There is nothing like a little humor to spice up a JAD session. After going through 52 data element definitions and 15 screen formats, people are ready for anything of a jovial nature that might give them some relief.

When the opportunity arises and you feel comfortable with it, try comic relief. For example, one Sunday when we were setting up the room for the session, the two-year-old daughter of the JAD leader happened to be with us. While we arranged magnetics on the board, she amused herself by making colorful drawings on blank flip chart paper. She made one particular drawing that we thought looked rather like a conceptual depiction of the application system we were designing (probably because both the drawing and the system were quite a mess). We labeled the drawing with some of the system buzzwords. We hung it in the front of the room so that it was the first thing everyone saw when they arrived. It was a good ice breaker. Throughout the session, people suggested new items to add to the picture. In the end, we had an impressionistic version of the complete system.

In another session, the group had developed such a good rapport that we felt it was appropriate to offer a round of champagne on the last day. To get everyone in a festive mood, we presented several humorous data elements, modeled after the real data elements. For example, we renamed the data element *Allocation Percent* to *Alcohol Percent*. *Original Issue Age* became *Original Drinking Age* and *IDB* (for Increasing Death Benefit) became *LCB* (for the Liquor Control Board). It took about five minutes to walk through them. The participants seemed to enjoy this, especially when we came to the part where the champagne was uncorked. We are certainly not recommending the bubbly stuff for all sessions. In fact, we have done it only that one time. But sometimes, going off the beaten JAD track in one way or another can be worthwhile.

For another JAD, we made up a humorous evaluation form modeled after the real one. We distributed it as though it were the real thing. Figure 8-1 shows this form.

In the same vein, there is no harm in applying a little humor when things go wrong. The group can appreciate certain events that might have occurred unbeknownst to them. One time, for example, we walked in half

---

JAD EVALUATION FORM

Name (optional) _____ Date _____

JAD Project _____

1. What was your participation in this JAD?

   _____ Bored user   _____ Bored MIS-user   _____ Bored observer

2. How would you rate your overall experience with JAD?

   _____ Very good   _____ Excellent   _____ Very excellent

3. How would you rate the overheads?

   _____ Crooked   _____ Out of focus   _____ Out of context

4. How would you rate the JAD leader's penmanship?

   _____ Looks like chicken scratch

   _____ Looks like hieroglyphics

   _____ Good for prescriptions but not for specifications

5. What color board markers do you prefer?

   _____ Magenta   _____ Ebony   _____ Chartreuse   _____ Lapis-lazuli

6. Do the following events have any connection?
   The Spanish Inquisition / The Reign of Terror / This JAD Session

   _____ Yes   _____ Yes

7. Do the following people have any connection?
   Torquemada / Robespierre / Denise Silver
   _____ Yes   _____ No   _____ Maybe

8. On the back of this form, please describe how JAD has affected your personal life.

Figure 8-1 Humorous evaluation form

an hour before the session and found our square table arrangement replaced with one hundred chairs all facing the same direction. As quickly as stage hands changing the set between scenes, we had to completely rearrange the room.

Another time, we arrived at the session an hour early to arrange the many magnetics and diagrams on the board. Once everything was set, we thought we'd relax a bit before the session. We went to the back of the room for some refreshments. Leisurely sipping some freshly brewed coffee, we glanced up at the board and realized the magnetics could not be read from the far side of the room. We had to switch gears and throw ourselves into reprinting all 54 magnetics in record time. We mentioned these war stories at the appropriate time in the session and people got a good chuckle.

These things are going to happen. When it feels right, you can turn your scars into stars (or at least turn your blunders into laughs) that lighten the mood, even for a moment.

Humor works best once group identity has had a chance to set in. In other words, you don't want to start right in with a preplanned joke about the "funny thing that happened to you on the way to the JAD session." Just let it come naturally and take the opportunities when they arise.

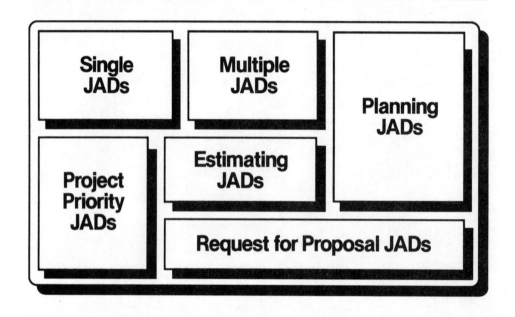

Single JADs

Multiple JADs

Planning JADs

Project Priority JADs

Estimating JADs

Request for Proposal JADs

NINE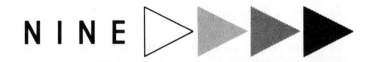

KINDS OF JADS

JAD projects come in many shapes and sizes. The scope of the project and the type of specifications you need determine how that particular JAD will be handled. Although the variations on a JAD are innumerable, we have found that they fall into the following categories:

- Single JADs
- Multiple JADs
- Planning JADs
- Project priority JADs
- Estimating JADs
- Request for Proposal (RFP) JADs

As you use the methodology, you may find other variations that work in your environment.

THE SINGLE JAD

Single JADs are like the one we have been describing throughout the book. They deal with a project that can be handled in one three- to five-day session. Usually they involve designing or enhancing one major application system. Typical single JADs that we have held include:

- designing an administrative system to offer existing policyholders the opportunity to receive higher policy dividends if they repay their policy loans.

- enhancing an underwriting and insurance policy issue system to support both individual and variable life products. This involved building a bridge between two computer systems, allowing transactions from one system to automatically pass to the other.

- designing a system to process insurance claims for one of our subsidiaries, a direct marketing insurance company.

If you have not used the JAD methodology before, begin with a single JAD for your first project. This is the simplest, most straightforward, and most predictable of all the JAD types.

MULTIPLE JADS

Multiple JADs are related JAD projects, usually scheduled back to back. They involve a series of interviews, sessions, more interviews, and more sessions. Furthermore, if you are working under tight time constraints, the phases of one JAD project can overlap the phases of the next. In the same week, you might be interviewing for one JAD and preparing the final document for another. There is much more involved in this approach.

So why even do overlapping projects if they are more complex and, as you can imagine, probably a bit intense? In fact, why not just schedule the JADs concurrently, using separate leaders? These JADs are usually held one after the other for the following reasons:

- you often need the output of one session to use in the next.

- the same participants are often required for several sessions. Some may even be required for all of them.

And why not spread the project over a longer period? The answer lies in the facts of the real world: When upper management says the system must be running by January 1, it *has* to be running by January 1.

In our organization, multiple JADs are held for enhancing systems to handle new insurance product introductions. To do this, we plan separate JADs for each system that must be modified. In other words, people attend only the sessions that affect their area. Each JAD has its own cast of characters, participation lists, and agenda.

Examples of our multiple JADs include:

- revising five systems to handle a new insurance product introduction that affected processing for underwriting and policy issue, batch and online policy administration, the agency network, commissions, and dividends

- a new disability income product introduction that, likewise, affected all the systems mentioned above

- revising five systems to handle the new tax laws passed by Congress

When you are not dealing with the constraint of an imposed deadline, you can schedule the JADs one at a time rather than weaving them together. Begin the second JAD when you have completed the session for the first. If you have more than one JAD leader, one person can compile the document for the first JAD while the other begins interviews for the second. This is the best way to handle multiple JADs.

PLANNING JADS

When you embark on a major project that will necessarily require several JADs, you can hold a planning JAD to identify how the project will be organized. Some major projects are clearly delineated by the computer system. Our multiple JAD projects were of this variety. In these cases, a planning JAD was not necessary. Other multiple JAD projects might be broken out along the lines of business functions. For example, the design effort for an Order Processing system might be organized into the following JAD projects:

- Customer order servicing
- Inventory management
- Shipping
- Receiving
- Accounts payable and accounts receivable

You need to assemble the key people from the user areas along with someone from MIS who understands the general impact the project will have on all systems involved. Objectives of the planning JAD might include:

- Define the overall project objectives and priorities
- Divide the project into a series of separate, manageable JADs
- Identify problems with the current environment
- Create preliminary participation lists for each JAD
- Set a preliminary JAD schedule
- Identify open issues that must be resolved before the JADs begin

For this kind of JAD, be careful about the distinctions between the roles of JAD leader and MIS project manager. The JAD leader remains the facilitator, as always, and should not assimilate the project manager's responsibilities of tracking project progress after the JAD session. For example, projects will involve program design, coding, testing, and implementation tasks. These responsibilities belong with the Applications Development team. At the same time, you should not be involved in overseeing such user area functions as preparing marketing brochures, researching legal implications, and so on. It may seem obvious that these are not your responsibilities, but it is surprising how certain responsibilities can improperly fall to the JAD leader, simply because of the apparent leadership role and documentation skills the position assumes.

PROJECT PRIORITY JADS

This unique JAD breaks away from the traditional systems design projects that most JADs address. The objective here is to prioritize a group of projects and agree upon target dates for implementing each one.

We ran a project priority JAD to sort out 55 separate projects that the Individual Insurance Division users had requested of MIS. Instead of making magnetics for data elements and screen names, we made them for the project names. Then, using the JAD methodology, we guided the participants through the following:

- Prioritizing the projects
- Estimating person-months for each project
- Identifying the status of the specifications (complete, partial, or nonexistent) for each project
- Graphing the time lines for each project
- Determining target dates (if the specifications were complete)

The participants included representatives from each of the ten departments that had projects in the mix. To handle cases where participants disagreed on the priority of their projects, we had the executive sponsor attend to arbitrate such conflicts. He offered a broader company perspective.

The target dates were agreed upon where possible, a priority committee was established, and a PC application was designed to track these priorities as they changed over time.

This JAD was like no other. It involved no mainframe application design. In the beginning of this project, we questioned whether it met the JAD criteria. As we looked into it, however, we saw that the project fulfilled two important criteria:

- It involved users from more than one department (in fact, it involved ten departments).
- It had a high business priority.

Furthermore, the project could benefit from the overall JAD structure, including the use of an impartial leader and a scribe to document the decisions.

ESTIMATING JADS

Some companies, such as New York Life, use JAD to estimate the cost of software development projects. Before jumping into designing a major system, the users can better gauge the cost and the value of a project using JAD. This helps them determine how they can best use their data processing resources.

As the people from New York Life describe, "An Estimating JAD . . . can last anywhere from 1 to 2 days depending on the complexity of the request. . . . Business functions are identified as well as inputs and outputs of those business functions. If the project impacts existing systems, then those systems are discussed as well to determine the scope of the enhancements needed. Certain technical issues are addressed at this time; for example, security, projected growth of the system and hardware requirements." (New York Life, 1986)

The final document of an Estimating JAD is used by MIS to determine cost estimates for the projects.

REQUEST FOR PROPOSAL (RFP) JADS

Request for Proposal (RFP) JADs work well when the users want to purchase a software package from an outside vendor rather than develop the application in-house. The JAD identifies user requirements that provide a basis for selecting the package. The final document, then, is used to create the Request for Proposal (RFP) which is sent out to vendors. The vendors respond to this document with their own proposal of how they plan to meet the company's requirements along with a cost estimate.

The JADs address a different level of detail. The final document may include the same agenda items (work flow, screens, and reports), but the information in these sections is less detailed. For example, we held an RFP JAD for a Fixed Assets system. In the Screens section, a normal document would show screen flow, screen descriptions, and complete screen designs. For the RFP JAD, we showed only screen flow and descriptions. A package would already have the detailed screen designs that could be tailored to fit the user's needs.

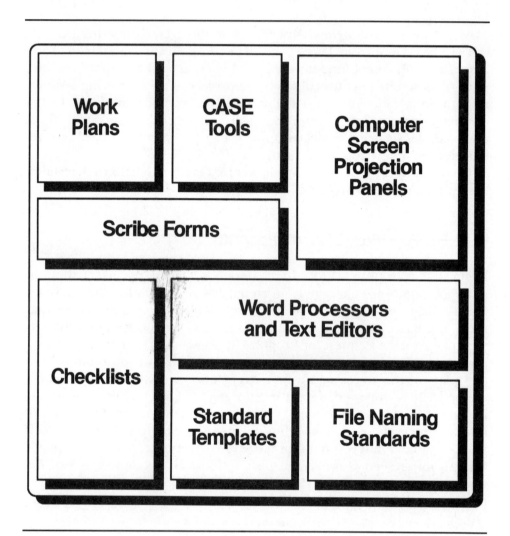

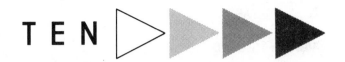
TOOLS AND TECHNIQUES

Every methodology has its own support tools and techniques—everything from handwritten forms to sophisticated software packages that require a 30-megabyte hard disk. The important thing to remember is that there is no one set of tools that guarantees a successful JAD. You do not have to use the exact same work plans, scribe forms, checklists, and other tools described in this chapter. We are offering examples of tools and techniques that have worked for us. You can use them as they are or modify them to fit your environment.

WORK PLANS THAT WORK

You can find information on how to plan projects from many sources. There are books, videos, self-paced courses, basic seminars, and advanced seminars. In addition, many software tools are available to help plan and track your projects. Just as in any major effort, JAD projects require some method for organizing and tracking the work. There are many ways to do this. The following describes one way—a technique we use to plan and track JAD projects.

There are two parts to this technique, one for *overall* planning (which schedules the JAD and its phases) and the other for *detail* planning (which plans and tracks individual tasks). The following describes each one.

Overall Planning

For single JADs, overall planning is nothing more than scheduling each JAD phase and finalizing the dates for the session. For multiple JADs, the planning is more complex. You are dealing with five phases for each project. Some of these phases may overlap and some of the participants may be part of more than one JAD. You cannot schedule to interview a user involved in one project when that user is attending a JAD session for another project.

For this kind of planning, we use a PC-based project management tool to create a PERT chart showing critical paths. Figure 10–1 shows a chart for a three-JAD project.

Detail Planning

For detail planning, we need an approach that is simple (easy to use and not requiring a lot of time to maintain) and flexible (allowing for constant changes and adjustments). We use an approach based on a series of *work*

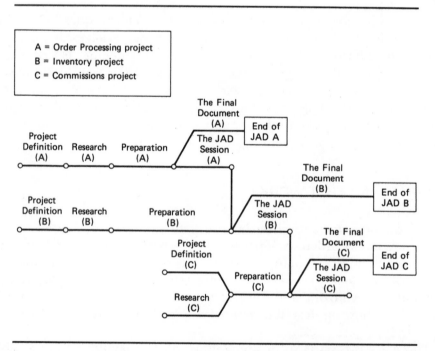

Figure 10-1 Overall plan for a multiple JAD project

Work Plan Order Processing System					
Task No.	Who	Task	Elapsed Time		
			Jan	Feb	Mar

Figure 10-2 Work plan form

plans that organize the tasks by showing when they should be completed and by whom.

The work plans are in a Gantt chart format (to track elapsed time), but they are not produced or maintained on a software package. Instead, they are based on forms that we fill in, review, and update daily. These can be done by hand or by using a word processor. Figure 10–2 shows a sample of this form. Beginning at the left side of the form, the columns are as follows:

- *Task No.* This is a sequential number assigned to each task to use when one task references another. For example, in the following work plan, task number 4 references task number 3:

Work Plan Order Processing System					
Task No.	Who	Task	Elapsed Time		
			Jan	Feb	Mar
3	Michael	Get list of existing screens from John			
4	Michael	Make overheads of screens from task no. 3			

- *Who.* This lists who will perform the task.

- *Task.* This is the actual work that needs to be done. Use clear descriptions that begin with verbs. For example, the following is *not* a good description:

Task No.	Who	Task	Elapsed Time Jan	Feb	Mar

Work Plan
Order Processing System

Task No.	Who	Task	Elapsed Time		
			Jan	Feb	Mar
2	Scott	Data elements			

This description does not tell what Scott should do with the data elements. Should he create a preliminary list of them, describe them for the Working Document, or write them on magnetics? A better description might be:

Work Plan
Order Processing System

Task No.	Who	Task	Elapsed Time		
			Jan	Feb	Mar
2	Scott	Create a preliminary list of existing data elements.			

- *Elapsed Time.* The three columns at the right side of the form show elapsed time. Fill in the headings with the time frames in whatever units you need. If you are planning over a period of months, the columns might be titled *Jan*, *Feb*, and *Mar*. For a shorter planning period, you could title the columns in weeks, such as *June 7, June 14*, and *June 21*. In other cases, you may not need to identify the time lines at all. For example, a week before the session, when it seems there are a thousand things to do, putting them down on this work plan helps identify all remaining tasks. Since you know that everything must be done this week (the session starts next Monday), you can simply note the tasks, determine who will do them, and delete them as they are completed.

Sometimes work plans can become unwieldy because there are so many tasks. In this case, divide the tasks into logical groups. You could, for example, make separate work plans for each JAD phase. The work plan for the Preparation phase might look like this:

Task No.	Who	Task	Elapsed Time Jan	Feb	Mar
		Work Plan Order Processing System Preparation Phase			
1		Review the inventory checklist			
2		Order supplies			
3		Prepare the Working Document			
4		Meet with the scribe			
5		Schedule the pre-JAD meeting			
6		Send pre-JAD meeting memo			
7		Send out Working Document			
8		Prepare flip charts — session agenda — system overview — open issues			
9		Prepare magnetics — data elements			
10		Prepare overheads — work flow — screens — reports			
11		Assemble scribe forms			
12		Arrange getting into the room before the session			

Simplicity Works Best

As you can see, this work plan technique is not a sophisticated processing tool that produces slick printouts. But this simplicity is exactly why the technique works for us. It provides an uncomplicated way of planning the tasks and tracking the progress of those tasks.

Tracking the Work

So far, we have talked about *planning* the work. The second part of project management is *tracking* the work to assure everything gets done in time.

For most JAD projects, you can simply track progress by reviewing the work plans each morning. Identify which tasks are done and cross them off the list. Evaluate the others to see if they are on schedule. And add any new tasks that have come up. A portion of such a work plan might look like this:

Work Plan
Order Processing System

Task No.	Who	Task	Elapsed Time		
			Jan	Feb	Mar
10	Susan	Prepare magnetics			
11	Sam	~~Check on accessing the meeting room~~			
12	Sam	Do flip charts			
13	Sam	~~Arrange for transporting JAD supplies~~			

If the tasks are many, you might want to track elapsed time both for when the work is planned and when it is performed. This is especially important when you are dealing with tasks that are dependent upon the completion of other tasks. To do this, use two elapsed time lines, one for planned work and one for completed work. In the following example, Lewis has completed task 5 and Gail is about halfway through task 6.

Work Plan
Order Processing System

————— = planned
XXXXX = completed

Task No.	Who	Task	Elapsed Time		
			8/1	8/8	8/15
5	Lewis	Print out all the Accounts Payable reports and give them to Sam	———— XXXX		
6	Gail	Evaluate which columns on the Accounts Payable reports are no longer needed		———— XXX	—

CAN CASE TOOLS HELP?

CASE (Computer Assisted Software Engineering) refers to software tools (usually PC-based) that support various parts of the systems development life cycle. Some examples are:

- Index Technology's Excelerator®
- KnowledgeWare's Information Engineering Workbench®
- Nastec's Design Aid®
- Yourdon's Analyst/Designer Toolkit®

Like JAD, these tools can increase development productivity and reduce maintenance. There is front-end CASE ("upper CASE"), which addresses the initiation, analysis, and design phases. And there is back-end CASE ("lower CASE"), which addresses the construction, implementation, and maintenance phases. An extremely sophisticated CASE tool would handle the entire systems development life cycle. However, most handle one particular part of that cycle.

How Are CASE Tools Used?

Here are some examples of how companies are using CASE:

- *Graphics.* This feature allows you to draw data flow diagrams, structure charts, data modeling diagrams, and entity relationship diagrams. All of these can help you develop and document system specifications. This is the most developed component of CASE.

- *Data dictionary.* The data dictionary is the core of many CASE tools around which all other features revolve. It is a central repository that contains definitions for all entities in the system being designed. For example, when you are creating data flow diagrams, you can define the data elements that make up the flows. This information is stored in the data dictionary and the definitions remain connected with the data elements. Then, when you design screens using those same data elements, their definitions are reflected in the screen. If the data element is defined to the dictionary as 10 characters long, it displays as a 10-character field on the screen.

- *Prototyping.* This allows you to "paint" proposed screens and reports. Then you can simulate menu branching where, for example, you make a selection on the main menu and the next screen displays as it would in the final system. (There is no actual code behind the screens, just the screen images.) Then you can enter data into the screen fields. Based on edits you have defined in the dictionary, the appropriate error messages display.

- *Code generation.* This creates code based on detailed specifications that you define. For example, one tool on the market translates your screen design into COBOL, BASIC, C, or PL/1 code.

- *Quality assurance.* This allows you to analyze what you have created with the CASE tool and detect errors in the design. For example, some packages analyze data flow diagrams for completeness (are all parts labeled?), syntax errors (such as do all processes have at least one data flow coming in and one going out?), and balancing (are the inputs and outputs of one data flow diagram equivalent to those on the next level?).

- *Other functions.* Some CASE tools support project management, project estimating, and word processing. These functions may be available within the CASE tool itself or through interfaces between the CASE tool and packages that perform these functions.

Which Comes First, the Methodology or the Tool?

Should you develop your methodology first, then purchase a CASE tool which supports that methodology, or should you develop your methodology afterwards based on the tool? For example:

- Structured analysis and design has been a way of life in your company for years. Programmer/analysts create data flow diagrams and structure charts following rigid standards. And now you want to buy a CASE tool to support this methodology.
- Or, on the other hand, each programmer/analyst works in his or her own homegrown style. And now you want to buy a CASE tool to force them to use a single methodology.

With CASE, the methodology should come first. Select the tools to support the way you work. If you bring in the tools first, then those tools will drive a methodology that may not be consistent with your original objectives. There is no point in purchasing a state-of-the-art tool just to get on the CASE bandwagon if you do not use the techniques offered. A tool that creates the most elegant data flow diagrams has no value if you don't use those kinds of diagrams.

To some degree, you can tailor the tool to your methodology. For example, you can select Yourdon diagramming techniques over Gane and Sarson. But you cannot tailor the tool to the overall way you develop systems.

CASE Tools and JAD

CASE tools can be useful in JAD, although they are by no means required. We ran many successful JADs before having any kind of CASE tool at all. Now, we use Excelerator, which creates data flow diagrams (which we use a lot for defining work flow), does quality assurance (to detect errors in those data flow diagrams), and does prototypes of screens and reports (which we use less often).

The best approach is this: If you are already using CASE tools, then incorporate them into your JAD methodology. For example, if you have a tool that creates data flow diagrams, use it in the work flow portion of the JAD. Otherwise (if you are not using CASE tools), set up your JAD methodology first. Then, add CASE tools for the portions that the tools can support. First the methodology, then the tools.

COMPUTER SCREEN PROJECTION PANELS

Computer screen projection panels display images from a PC onto a surface large enough to be viewed by the entire group of participants. These portable units (small enough to fit into a briefcase) are replacing the larger, more expensive projection systems (sometimes suspended from ceilings).

The projection panel sits on the flat part of the overhead projector and interfaces with the PC. Through liquid crystal display technology, it displays the PC images onto a large screen or wall.

This tool is useful if, for example, you want to change screen images right there in the session through whatever software package you have on the PC. Revisions can be made on the spot. You can also use this tool to display and update data flow diagrams, data element definitions, and report designs stored in a CASE tool dictionary.

A warning: If you are going to use any kind of PC tool to display and update images in the session, make sure you can make those changes smoothly and quickly. You do not want to hold up the session while you plod through nested menus, stumble through move and copy commands, and apologize for slow response time. If this is the case, you are much better off *not* using the PC and projecting them using traditional overhead transparencies instead.

SCRIBE FORMS

Throughout the book, we have referred to scribe forms that the scribe uses to record agreements made in the JAD session. The information that goes on these forms is used to build the final document. Therefore, the forms are designed in the same format as the pages of the final document. For example, the Open Issue scribe form has lines for the name of the open issue, whom it is assigned to, the date it will be resolved by, and the description. The Open Issues pages of the final document have exactly the same entries.

The original IBM JAD methodology includes scribe forms that look quite different from the ones we use. We have modified the forms and added several more to handle our needs. Although we have many scribe forms, we use only 5 to 10 types of forms per JAD project. Also, we usually make one or two custom scribe forms for each JAD. For example, one project involved identifying a series of transactions that the system would require. For each transaction, we documented transaction code, description, and so on. We designed a scribe form listing all these items. As it turned out, the group identified 80 transactions, each of which was recorded on a separate scribe form. The scribe's task of recording all these transactions was simplified and the possibility of omitting information was reduced.

When Should You Use Scribe Forms?

When updating existing information from the Working Document, you do not need a scribe form. For example, if you are changing the wording on an assumption already contained in the Working Document, the scribe notes the change on his or her copy of the document.

On the other hand, all new information generated during the session must be recorded on scribe forms. New information comes in two forms: repetitive and free form. Data element definitions are an example of repetitive information. You capture the data element name, length, format, and description for one data element, then another, and so on. For each data element, the scribe uses a separate scribe form with the heading *Data Elements* and labeled lines for *data element name, length, format,* and *description.* Free form information is handled differently. Screen Flow, for example, requires drawing one large hierarchical diagram that might spread across several blank pages. The scribe form, therefore, includes only the heading *Screen Flow* and nothing else.

You might wonder, why have a special form that contains nothing more than a heading? The standard heading will help you sort out which notes go with which agenda items after the session. You will avoid situations where you wind up lamenting, "Oh yes, I remember talking about this, but when was it? Where exactly does it fit in? It all seemed so clear in the session."

Bringing the Forms to the Session

For each item on the agenda, the scribe uses at least one kind of scribe form and sometimes several. When covering screens, for example, the scribe might use forms for:

- screen descriptions (to identify the required screens and their functions)
- screen flow (to show the hierarchy of how the screens branch)
- screen design (to show how the screen is formatted)

Furthermore, for each type of scribe form, several copies can be used. When designing screens, for example, the scribe uses a separate form for each new screen design.

You can imagine the large quantity and variety of forms you need to have at the session and the importance of keeping them well organized. Sometimes, while the participants are thinking and analyzing, the scribe records nothing for what seems like an eternity. Other times, the increased pace has the scribe hopping from form to form. When designing screens, for example, new data elements arise, old ones change, open issues accumulate, and assumptions are made. The scribe needs to get hold of the right form with no delay.

To assure that the scribe can do this, you can put the forms in separate hanging folders labeled by their headings (such as Assumptions, Data Elements, and Open Issues). These folders are sorted alphabetically and put into the high-tech state-of-the-art filing device known as a cardboard box (with the lid cut off). The scribe either keeps the box on a nearby chair or arranges the more frequently used forms in piles on the table. When an agreement is made, he or she takes out a form and fills it in. The scribe arranges the completed forms on the table for easy accessibility—for example, if the group wants to change a data element or read back an assumption. The scribe can arrange the forms by whatever method works best. Some scribes put them in order of the agenda items, others in alphabetical order by form name. Any method is fine, as long as the forms can be reached quickly.

How many forms should you bring? The answer is, bring enough so you don't run out. Bring twice as many as you think you need. You can always use them in the next JAD. We bring 10 to 75 copies of each form depending on their use. The forms that require the most copies (and which we never seem to have enough of) are:

- Assumption
- Open Issue
- Data Element Description

Scribe Form Samples

Appendixes A and B show samples of the most common scribe forms that we use. You can copy them directly from the book or modify them for your needs. The actual forms are on 8–1/2-by-11 inch paper. Therefore, if you copy them, enlarge the image to fit that size.

These forms are divided into two groups:

- forms for the Management Definition Guide
- forms for the JAD session

The following describes each group.

Forms for the Management Definition Guide

The first group of forms, shown in Appendix A, is for those initial interviews when you are gathering information for the Management Definition Guide. The "Interviewing Management" section of Chapter 3 describes how to use these forms. The form names are:

- Purpose of the System
- Scope of the System
- Management Objectives
- Functions
- Constraints
- Additional User Resource Requirements
- Assumption (Pre-Session Version)
- Open Issue (Pre-Session Version)
- Participation List

Appendix A shows samples of these forms.

Forms for the JAD Session

Samples of these forms are shown in Appendix B. Figure 10–3 lists each agenda item and which forms support it. The last two columns show whether the form is standard (used for many JADs) or custom (designed for a specific JAD). Remember, you will probably need only 5 to 10 types of forms per project.

CHECKLISTS

Over the course of our JAD projects, we have developed a few checklists that have helped along the way. These include checklists for:

Agenda Item	Form Name	Standard Form	Custom Form
Common Scribe Forms			
Assumptions	Assumption	X	
Work Flow	Work Flow Diagram	X	
Data Elements	Data Element Description	X	
Screens	Screen Flow Diagram	X	
	Screen Description	X	
	Screen Design	X	
	Screen Message	X	
	Fields by Screen		X
	Screen Access by Job Function	X	
Reports	Report Description	X	
	Report Design	X	
	Report List		X
Open Issues	Open Issue	X	
Other Scribe Forms			
Records	Record Description		X
	Record Design		X
	Record Volume		X
	Set Length		X
	Logical Design		X
Transactions	Transaction		X
Processing	Calculation Routine	X	
Manual Forms	Manual Form Description	X	
	Manual Form Design	X	

Figure 10-3 Scribe forms for the JAD session

- JAD tasks
- JAD supplies
- scribe forms

The next few pages show samples of these lists and describe how to use them.

JAD Tasks Checklist

This checklist shows all the tasks that could be required for doing a JAD. It is the basis for creating work plans as described in the "Work Plans That Work" section of this chapter. You can use this checklist as a starting point. The last column, called *Required?*, allows you to note whether you need to do that particular task for this project. By checking the required

JAD Tasks Checklist

Task	Required?
Phase 1: Project Definition	
Interview executive sponsor Interview users Interview MIS Create the participation list Prepare the Management Definition Guide Schedule the session Reserve off-site meeting room Send JAD session cover memo to all participants Take inventory of JAD supplies	
Phase 2: Research	
Interview users for system familiarization Interview MIS for system familiarization Define and document work flow Gather preliminary data element, screen, and report definitions Prepare session agenda	
Phase 3: Preparation	
Schedule the pre-JAD session meeting Send memo on pre-JAD session meeting Prepare the Working Document Prepare the script for the session Meet with the scribe Prepare flip charts, magnetics, and overheads Hold pre-JAD session meeting Set up the session meeting room	
Phase 4: The JAD Session	
Follow the script	
Phase 5: The Final Document	
Schedule the review meeting Prepare draft of the final document Distribute the final document and memo Hold the review meeting Update and distribute the document, if necessary Get signatures on approval form	

Figure 10-4 JAD tasks checklist

tasks, you will know what to include in the work plans. Figure 10–4 shows a sample of this checklist.

JAD Supplies Checklist

You do not want to run out of certain supplies during a JAD session. If you run out of rubber bands, staples, or No. 2 pencils, you can almost certainly scrounge some up around the department. But let's say that on the day before the session you need to make overhead transparencies for five pages of work flow, eight proposed screen designs, and ten existing reports. You go to the copy machine and, lo and behold, there are no blank transparencies. You ask the secretary for some more. She tells you she gave the last ones away yesterday, but they are on order and will be in next week.

To avoid this unnecessary stress, take an inventory of supplies as soon as the project begins. At that time, determine what you need, check what you have, and order what you don't have. Figure 10-5 shows the JAD supplies checklist that we use to check our inventory. This checklist helps you to track items on order as well as remember what to bring to the session.

Scribe Forms Checklist

Part of preparing for the session is determining which scribe forms to use. To do this, review the Working Document and JAD agenda to identify the topics to be covered along with their corresponding forms.

To help select these forms, you can use the scribe forms checklist shown in Figure 10-6. It lists all available forms and has the following columns:

- *Agenda Item*. This tells where in the agenda the form is used.
- *Form Name*. This lists the name of the form as shown in the form heading.
- *Required?* This is a blank column you can use to either mark with a check or note the number of forms to bring to the session.

WORD PROCESSORS AND TEXT EDITORS

What is the best way to produce your JAD documents? Should you use a word processor or a mainframe text editor (assuming you have a choice)?

A word processor runs on a personal computer using one of the many word processing packages available, such as:

- pfs:write®
- Multimate®
- WordPerfect®

The features of these packages vary greatly, so the effectiveness of using a word processor depends a lot on which package you use.

A text editor, on the other hand, runs on a mainframe computer. Users access the text editor via a terminal such as the IBM 3270-type series. We use IBM's Document Composition Facility (Script/VS®) combined with Image Science's DCF/PLUS®(which allows Script/VS to interface with the Xerox laser printer).

JAD Supplies Checklist

Need to Order	Date Ordered	Date Received	Item
Overheads			
			Bulb and spare
			Blank transparencies
			Yellow transparency
			Transparency frames
			Vis-a-Vis pens
Magnetics			
			1.5" by 8" shapes
			5" by 7" shapes
			3" by 10" shapes
			CRT shapes
			Disk file shapes
Flip Charts			
			Flip chart paper
			Pens
Final Document Assembly			
			Binders
			Tabs
			Cover, back, & spine
Other Supplies			
			Board pens
			Erasers
			Eraser juice
			Pencils
			Tablets
			Tents
			Masking tape
			Scotch tape
			Stapler
			Staples
			Scissors
			Rubber bands
			Paper clips
			Company telephone book

Figure 10-5 JAD supplies checklist

Scribe Forms Checklist

Agenda Item	Form Name	Required?
	Common Scribe Forms	
Assumptions	Assumption	
Work Flow	Work Flow Diagram	
Data Elements	Data Element Description	
Screens	Screen Flow Diagram Screen Description Screen Design Screen Message Fields by Screen Screen Access by Job Function	
Reports	Report Description Report Design Report List	
Open Issues	Open Issue	
	Other Scribe Forms	
Records	Record Description Record Design Record Volume Set Length Logical Design	
Transactions	Transaction	
Processing	Calculation Routine	
Manual Forms	Manual Form Description Manual Form Design	

Figure 10-6 Scribe forms checklist

Comparing Word Processors and Text Editors

To make your decision, you need to evaluate how each approach meets the criteria you have set. The following lists some possible criteria and describes how each approach meets those criteria.

Turnaround Time

A PC word processor gives you better turnaround time than a mainframe text editor. Getting your printouts is simply a matter of selecting the print option. The pages are produced from a printer directly connected to that PC or one nearby.

Using a mainframe takes more time. When you order the printout, the request queues up with other print requests. It usually prints in another location (sometimes in a different building) where the printouts are sorted and distributed. This can take anywhere from half an hour to half a day or more, depending on the work load in the computer room (for example, are the company's payroll checks printing?), the status of the printer (it could be on the blink), and how often your company distributes printouts.

What You See versus What You Get

With a word processor, the text you enter displays on the screen in much the same format as it will on the printout. What you see is what you get. Word processors have automatic word wrap, which means that you enter text in a continuous stream without having to determine where a line begins and ends. This enables you to enter the text more rapidly.

With a text editor, what you see is not exactly what you get. As you enter text, the screen displays not only the text, but also line numbers and codes. Automatic word wrap does not happen as you type. The commands for things like spacing and tabs are entered as codes, which are translated into the format you want. For example, you enter a code such as .sp 3, which, after processing, causes three lines to be skipped in the final text.

Ease of Use

Word processors are easier to learn and easier to use. Text editors require learning many codes (like a whole separate language), and because of these codes are somewhat more complicated to use.

Power

By now you may be wondering, why would anyone use a mainframe text editor that has slower turnaround time and forces you to mix cumbersome codes with your text? Well, here is where text editors excel—in the *power* you have in formatting, paging, indexing, and other functions.

Usually, text editors allow you to automatically generate tables of contents, lists of illustrations, and indexes. For the heading of this chapter, "Tools and Techniques," for example, you might type ":h2.Tools and Techniques." The text editor automatically puts that heading in the table of contents and calculates the page number where it appears. A text editor might handle figures in the same way. For Figure 10-7, a text editor would automatically assign the figure number and include it in a list of illustrations. Indexes can be generated as well. These features are valuable when producing the Management Definition Guide, the Working Document, and the final JAD Design Document.

Criteria	Word Processors	Text Editors
Turnaround time	best	
Seeing what you get	best	
Ease of use	best	
Power		best
Data protection		best

Figure 10-7 Word processors versus text editors

Also, text editors give you more control over such things as highlighting, indentation, boxes, and font changes. And they are great with lists, such as:

1. item one

2. item two

3. item three

The items in such a list are automatically numbered so that if you add one item (for example, between items one and two), you do not have to change all the other numbers.

Data Protection

Here is another area where text editors have an advantage. Data is less likely to be lost using a text editor. Furthermore, when you do lose data, you will more likely get it back.

All of us have experienced losing hours or days worth of work. Perhaps a power failure caused all your data to be lost just before you had a chance to back it up on a floppy. Or maybe your toddler pulled out the PC plug just to get your attention. Using a PC means that you are responsible for backing up your own work. Using a mainframe, your work is probably backed up on a regular schedule, every night for example. Those backups may be taken off-site and archived as well. You do not have to think about it until, of course, you lose your data. Then the work can be recovered from the last time it was backed up.

Which Approach Is Best for You?

Although word processors are easier to learn and use, text editors offer more power and data protection. Figure 10–7 evaluates the five criteria we have covered. "Best" in the column indicates which approach best meets the criteria.

So, which approach should you use? That depends on which criteria are important to you. For example, if there is a lot of turnover in the people creating the documents, you might want to use a word processor that is easier to learn and use. On the other hand, if the same person always produces the documents, then the learning curve does not have to be repeated. Once that person learns the text editor, more power and better data protection are available.

STANDARD TEMPLATES

Standard templates are skeletal files used to create other files. For example, a Human Resources department could have standard templates for their letters offering employment. "Dear xxxx," might begin the letter. When an offer is made, the secretary copies that standard template and replaces the "xxxx" with the appropriate name. In the same way, you can use standard templates for the Management Definition Guide, the Working Document, and the final JAD Design Document. Let's say, for example, you gathered 20 data element definitions before the JAD session and you want to include them in the Working Document. For each data element, you need the following information:

- name
- length
- format
- description

Rather than type the same labels for that information 20 times (once for each data element), you can copy one file with the base information (this is the standard template), then add all the variable information. The standard template might look like this:

DATA ELEMENT DESCRIPTIONS

Name: xxxx

Length: xxxx

Format: xxxx

Description: xxxx

After you copy the standard template and replace the xxxx's with the information for that data element, the final version might look like this:

DATA ELEMENT DESCRIPTIONS

Name: Customer Number

Length: 7

Format: Numeric

Description: A unique number assigned to each customer.

After the session, when you produce the final document, you can use standard templates in the same way. You enter the information from the scribe forms into templates that are in the same format as the pages in the final document.

We have standard templates for all sections of the final document and, consequently, the Working Document, since they are both in the same format. Each template includes headings, index entries, and the standard format for the contents of that section. You can keep templates for the title page, preface, agenda, assumptions, data elements, screens, reports, and open issues. You can also keep templates for the following memos:

- to announce the pre-JAD session meeting
- to provide meeting information (date, time, location, and directions) for the JAD session
- to send with the Working Document
- to send with the draft of the final document and to announce the review meeting
- to send with the final JAD Design Document
- to send with the approval form

Templates are also useful for forms and checklists such as the following:

- document distribution form
- work plan forms
- JAD tasks checklist
- JAD supplies checklist
- scribe forms checklist

Using standard templates is obviously a great timesaver in the document creation process. Although you may spend a lot of time setting up templates in the beginning, once you have them, you can produce comprehensive final documents in a very short time. Forms and checklists are immediately available. And preparing memos involves nothing more than copying the standard templates and replacing the variable information.

FILE NAMING STANDARDS

To use standard templates most effectively, you should establish file naming standards. In other words, organize your documents into separate computer files and name them in a logical way.

We have set up standards for naming files for the Working Document and the final JAD Design Document. Following is a description of how we do it.

Naming Document Sections

You can use a seven-character file name, where the first four characters indicate the document section and the last three indicate where in that section the file falls. For example, if you are preparing the screens section, make a separate file for each screen. The first four characters are *scrn*, for screen. The last three are *010*, *020*, *030*, and so on. Therefore, the screen section might include:

scrn010	Add Orders
scrn020	Change Orders
scrn030	Print Orders
scrn040	Print Packing List
and so on	

The numbers are incremented by 10 so that you can insert new files between existing files. In the above example, you can add *scrn015* between *scrn010* and *scrn020*.

If you need to use two files for one screen (for example, one for the existing screen and the other for the new screen), add an eighth character. Add *e* for existing and *n* for new. Now, the screen names are:

scrn010e	Add Orders (existing)
scrn010n	Add Orders (new)
scrn020e	Change Orders (existing)
scrn020n	Change Orders (new)
and so on	

We use the following section codes:

```
over    JAD overview
agen    agenda
assm    assumptions
work    work flow
elem    data elements
scrn    screens
rept    reports
issu    issues
```

Naming Section Introductions

To name the introductions for these sections, we use two zeros at the end of the file name. For example:

```
scrn00     Introduction to Screens
rept00     Introduction to Reports
```

Benefits of File Naming Standards

Using these file naming standards helps keep your document parts in a meaningful sequence. When you need to access a file from a past document, you will be able to find it quickly. If, for example, you want to look at the fourth report in a document produced two years ago, the standards tell you that the file name is rept040.

File naming standards also help when one person takes over work for another. If the new person has worked on any JAD documents before, the file names will be familiar and easy to use. The person can step right into the project and be productive.

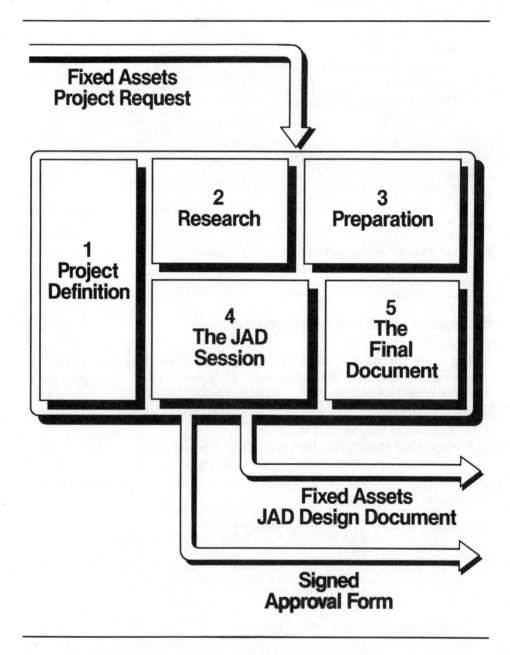

Fixed Assets
Project Request

2
Research

3
Preparation

1
Project
Definition

4
The JAD
Session

5
The
Final
Document

Fixed Assets
JAD Design Document

Signed
Approval Form

ELEVEN ▷ ▶ ▶ ▶

CASE STUDY: A REAL LIVE JAD

This chapter follows from start to finish an actual JAD project that we conducted. It covers some of the approaches we used and, more importantly, it covers the problems, stumbling blocks, and political dilemmas that we encountered along the way. To get the most out of this chapter, we recommend reading or scanning:

- Appendix C (the Management Definition Guide)
- Appendix D (the final JAD Design Document)

DESIGNING A FIXED ASSETS SYSTEM

The project we describe here is the design of a Fixed Assets system. We cover it from start to finish and include samples of the Management Definition Guide and the final JAD Design Document.

This project makes a good case study because it is filled with exceptions (no two JADs are alike), successes (most parts worked very well), and blunders (will you still have confidence in us if we share with you our mistakes as well?). You can probably learn more from hearing about the parts that went wrong than hearing about how everything ran smoothly.

Our JADs usually involved major insurance applications. We were used to dealing with such terms as modal premiums, paid-up additions, and cash values—concepts rather specific to the insurance industry. The Fixed

Assets project involved such things as capital gains, depreciation, and amortization—matters common to virtually all businesses.

The Objective

Our task in this JAD project was to determine the requirements for a Fixed Assets system for a medium-sized insurance company.

"Fixed assets," in case the term is unfamiliar, are the distinct, physical pieces of equipment a company owns or uses in its business. Distinct from real estate and liquid assets (cash, bank deposits, stocks, bonds, and so on), fixed assets are things you can put your hands on—such items as desks, chairs, and computer terminals. In its simplest form, a *Fixed Assets system* is just a data base for storing where all this material is located. In addition, many companies want their Fixed Assets systems to track when they acquired each item, how much it cost, and how long its useful life is. Then they can calculate depreciation for tax purposes.

Our existing Fixed Assets system tracked assets for home office furniture, field furniture (in the agencies), and data processing (EDP) equipment. We needed a new system that would also track field equipment, home office equipment, and leasehold improvements (carpeting, electrical wiring, and water fountains). Since the company had acquired several subsidiaries, the new system needed to track their fixed assets as well.

The Circumstances

The Finance department needed to determine whether to purchase a software package from an outside vendor or have a system written in-house by the MIS department. Therefore, they did not need detailed specifications—at least not yet. They needed only enough information to send a Request for Proposal (RFP) to the potential vendors as well as to get an estimate from the in-house MIS project team as to how long the project would take and how much it would cost.

The Cast of Characters

Among the people involved in this project were:

- *The Executive Sponsor.* We selected the Vice President and Controller because she was the most senior person responsible for the key departments that appeared to be directly affected by the project. She was able to make the necessary management-level decisions throughout the JAD project.

- *The Scribe*. A programmer from our Applications Development group did the scribing. He would be the one programming the new system if it were designed in-house.
- *The Participants*. Because the project affected such a large number of departments, the participant list was larger than normal. Some of the 18 participants had not even met each other until this project. The complete list of participants is shown in the sample documents in Appendixes C and D.
- *The JAD Leaders*. Instead of having one JAD leader and a support person, we had two leaders (the authors). Each shared the leader and support responsibilities.

The Documents

We prepared all three JAD documents:

- Management Definition Guide
- Working Document
- JAD Design Document

Since this was an RFP JAD, the Working Document and the JAD Design Document contained all the standard parts of a final document except data element definitions and detailed screen and report design. Samples of these detailed items that do not appear in this case study are shown in Chapter 12.

Figure 11–1 shows a data flow diagram of the five JAD phases. The case study that follows describes the activities in each phase for the Fixed Assets project, and shows where that phase fits into this diagram.

PHASE 1: PROJECT DEFINITION

In this phase, we (the JAD leaders) interviewed management, selected the participants, prepared the Management Definition Guide, and scheduled the session. The following shows the Phase 1 segment of the diagram in Figure 11–1.

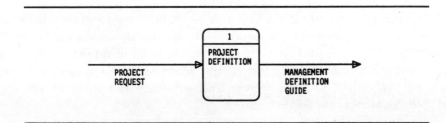

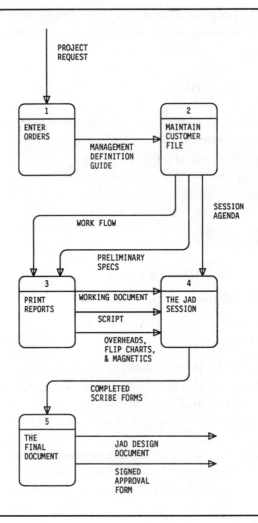

Figure 11-1 Diagram of the five JAD phases

Interviewing Management

In separate meetings, we met with the key users, the executive sponsor, and MIS.

Our objectives for the first key user meeting were to gather information for the Management Definition Guide, to identify at least some of the session participants, and to discuss when the session might be held. We began asking questions relating to the Management Definition Guide such as, "Who will use the system and what are the interfaces?" This defined the *Scope of the System* section:

SCOPE OF THE SYSTEM

The Fixed Assets system will include:

- Home Office Furniture
- Home Office Equipment
- Field Furniture
- Field Equipment
- EDP Equipment
- Leasehold Improvements
- Multi-Company Processing

The following areas of the company are affected either through direct system interaction or indirectly by providing source documents or receiving reports:

- Budget and Cost
- Facilities Administration
- ISD Planning
- Marketing Properties
- Purchasing
- Tax Planning

The Fixed Assets system will interface with:

- CDS (Cash Disbursement System) for acquisitions
- ABC (Accounting Budget and Cost) for depreciation, chargeback, and disposal accounting of Home Office Furniture; unit cost accounting and chargeback of Field Furniture
- AER (Automated Expense Redistribution) for EDP equipment chargeback

---

We asked, "What do you want to gain from the system?" This helped build the *Management Objectives* section, which included such items as:

- record the acquisition, depreciation, and disposition of fixed assets to conform with STAT/GAAP and Tax guidelines
- reconcile with the company's trial balance asset accounts

And we asked, "What actual functions must the system provide?" This is when we encountered the first obstacle. The participants were having a hard time answering the questions we were asking. Apparently, we had not made the meeting agenda clear enough. They did not realize the details we were after and had not had a chance to sit down together and figure out exactly what they wanted the Fixed Assets system to do. We closed the meeting and reconvened later, giving them time to research the questions.

Then, back at the office, we began to question whether we should even be holding a JAD for this RFP initiative. In retrospect, we can see that this doubting was due to our never having done this kind of JAD before. This project was not a traditional JAD endeavor. But in fact, it met all the JAD criteria and, in the end, was well worth doing. So, beware of this feeling of uncertainty. It can happen whenever you break away from the traditional mold of JAD projects. It will probably subside in a day or two, once the project gets rolling.

We reconvened the key user meeting and completed the questions for the Management Definition Guide. (After this project, we changed our approach. Now we give the users special forms to fill out—shown in Appendix A—to capture this information ahead of time rather than trying to extract it during meetings.)

After this second meeting, the *Functions* section was compiled to include such items as:

---

1. Add, dispose, and transfer inventory items
2. Display information
    a. Items by category ID
    b. Items by cost center
    c. Category table
    d. Item record
3. Update depreciation information
    a. Handle both tax and statutory information
    b. Calculate monthly depreciation
    c. Update item records
    d. Feed accounting entries to ABC
4. Calculate gain/loss for statutory and tax reporting

After the key user meetings, we were ready to meet with the executive sponsor. The executive sponsor was a company Vice President whom we had not met before. We felt we had to be especially prepared. In the meeting, we spent relatively little time on system functions and more time on the purpose, scope, and objectives of the project. She defined "fixed assets" as they related to our company. This helped define the *Purpose of the System* section, which went into the Management Definition Guide as follows:

PURPOSE OF THE SYSTEM

The purpose of the Fixed Assets system is to provide accounting and reporting information for the Finance department to track the capital assets purchased for the company and its subsidiaries. For this project, fixed assets includes those items that:

- have useful lives greater than 12 months
- satisfy capital requirements as determined by asset classification
- are intended to be held in the operating activity of the business
- are tangible, that is, have physical substance

She further defined exactly which departments would be affected by the system. This was added to the *Scope of the System* section previously shown. She resolved some open issues. For example, the question of whether data entry should be centralized in the Finance department or remain in separate departments was resolved into the following assumption:

Centralization in Finance

- For all fixed assets in the company, Finance will reconcile the detail records in the Purchasing ledger as well as handle payments. Purchase decisions will continue to be made in their related areas.
- All data entry (other than Information Services Operations) will be done in Finance for assets purchased by the company. Other departments using the current system will be able to inquire into the system. Tax Planning will be able to approve tax depreciation and useful life.
- The Finance departments of subsidiaries that purchase their own fixed assets will update their own fixed assets information.
- Leasehold improvements are currently handled on a Lotus 1-2-3 program for depreciation in the Mortgage Loan and Real Estate department. This will also move to Finance.
- Finance will be staffed to handle the new Fixed Assets system work flow. The needs will be analyzed after user requirements have been defined in the JAD session.

Finally, we met with our MIS department (called Information Services Department, or ISD for short). We focused on the technical problems with the existing system. (For example, how should the data base be reorganized to handle the new system design—if it were developed in-house?)

Selecting the Participants

During each interview (with the users, executive sponsor, and ISD), more names were added to the participation list. The list included people from Accounting, Finance, Budget and Cost, Tax Planning, Facilities Management, Administrative Services, and ISD.

This JAD offered an interesting twist because one of the ISD participants attended the session as a user. Since data processing equipment also must be tracked by the Fixed Assets system, someone from ISD was responsible for adding new equipment entries into the system, defining the depreciation amounts, and tracking the assets. Therefore, he was affected by the decisions and needed to attend the session.

Page 293 of Appendix C (the Management Definition Guide) shows the final participation list and the individuals' departments (although the real names have been changed to protect the innocent).

The Management Definition Guide

Based on the information gathered in the interviews, we compiled the Management Definition Guide. We distributed a draft copy to the key users and executive sponsor for their review. We made their recommended changes and distributed the final version to all the participants.

Appendix C shows the Management Definition Guide. We have abbreviated some of it. For example, rather than showing pages and pages of functions for all the fixed assets groupings, we included only home office furniture, field furniture, and multi-company processing. The remainder of this case study will be clearer if you take the time to read the Management Definition Guide at this point.

As you read the document, you may notice a flaw in the Management Objectives section. Most of the items are not objectives at all; they are rehashed functions that should have been a part of the Functions section. We have since fine-tuned our definitions for the Management Objectives section. For a description of how to define management objectives, see the "Management Definition Guide" section of Chapter 3.

Scheduling the Session

Based on the users' schedule for sending the RFPs to the vendors and the availability of the key users and executive sponsor, we set tentative dates for the session. We called Debbie, our contact at the off-site facility we often use, to set the date and times for the meeting rooms.

Since the session was to be held in October, we had difficulty getting a room. (Scheduling meeting rooms in autumn is always a challenge. Summer vacations are over so you are competing with back-to-school and back-to-work events.) We finally worked it out, but had to settle for alternating rooms within the same facility (one room on Monday, another on Tuesday, back to the first one on Wednesday, and so on.) At this time, we also gave Debbie the "Danish" count, the setup we wanted for the room, and the audio-visual requirements. (In this case, we needed only a projection screen and a small table for the overhead projector.) We sent the meeting information memo to all participants.

After interviewing management, selecting the participants, producing the Management Definition Guide, and scheduling the session, we were ready to move on to Phase 2: Research.

PHASE 2: RESEARCH

In this phase, we learned about the existing system, documented work flow, gathered preliminary specifications, and finalized the JAD session agenda.

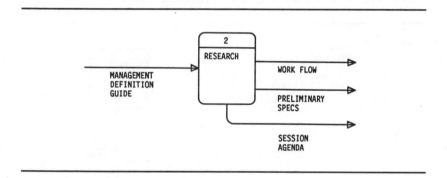

Getting Familiar with the System

To learn about the existing system, we spent some time with the users. They showed us how they used the system on a daily basis. We talked about what parts of the system worked well, and where the problems lay. For example, one user said that tracking fixed assets for home office furniture was difficult because the furniture was defined by entire workstation rather than by individual component part. There were "A" stations, "B" stations, and so on, depending on what furniture the workstation contained. If someone moved a filing cabinet from one station to another, there was no way to track it. Another user said that adding data processing equipment to the fixed assets data base was a problem because of how depreciation was handled.

We also held two brief meetings with the programmer/analyst, Jane, who had worked with the existing Fixed Assets system for years. Using a mainframe terminal, she walked us through the functions of the system, branching from screen to screen and making typical entries. She prepared hardcopy samples of all screens and reports. And she showed us a diagram of the data base design explaining the record types, relationships, and navigation routes through the data base.

Even though these meetings with users and ISD familiarized us with the existing system, we felt it would help to know how other insurance companies handle their fixed assets before we began defining requirements for a new system. So, through some connections of the Accounting department, we arranged a meeting with the Financial Services department of another insurance company. We spent half a day hearing about how their system worked, picking up some helpful hints, and generally commiserating about the "wonderful world" of fixed assets.

Documenting Work Flow

This part of the JAD ran almost exactly "by the book." To document the work flow, we used our PC-based CASE tool, Excelerator, to create the

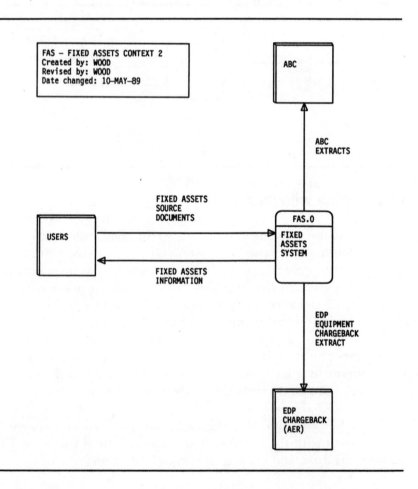

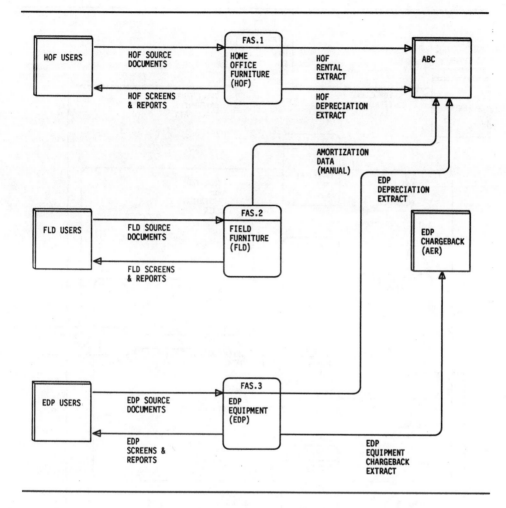

data flow diagrams. We held a meeting every morning for one week with the two people most familiar with the flow.

On Monday, we captured the first two levels of *existing* data flow, using the board, flip charts, and Post-it notes. The charts on pages 204 and 205 show the results of this meeting, which were included in the final JAD Design Document.

On Tuesday, we reviewed these two levels on the overhead projector and went on to identify the third and fourth more detailed levels. The chart on page 206 shows the third-level explosion of process FAS.3 (EDP Equipment).

On Wednesday, we reviewed all the levels and finalized them. Then, using these levels of existing data flow as a base, we defined how the *proposed* new work flow might work. The chart on page 207 shows the context diagram for the new work flow. It includes the added interface to the Cash Disbursement System (CDS).

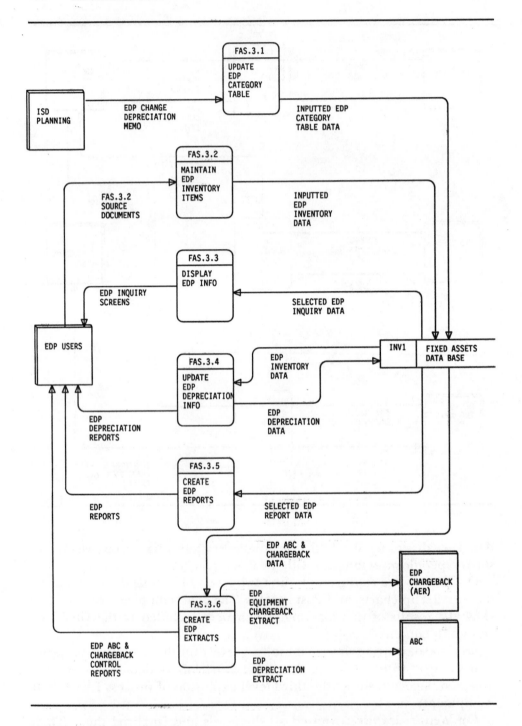

On Thursday, in a brief meeting, we reviewed the proposed new work flow. On Friday, we showed the entire work flow to another group of users. We added their recommendations to the diagrams.

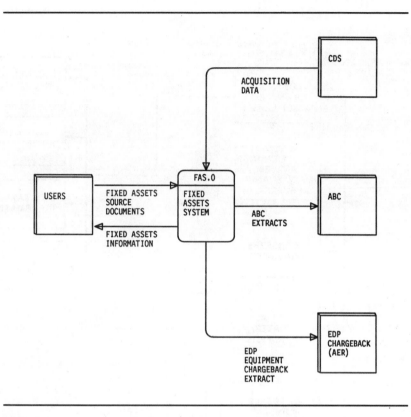

By the time the work flow diagrams were presented at the session, most of the group had participated in creating them. They understood the language of data flow diagrams. And they had some ownership, some vested interest in the documented work flow from having been directly involved.

All these diagrams were included in the Working Document as well as the final document. Pages 313–325 of Appendix D (the final document) show the complete data flow diagrams for the existing work flow.

Gathering Preliminary Specifications

Since this was an RFP JAD, we did not gather information for data elements. We did, however, prepare preliminary specifications for screens and reports.

For screen requirements, we were fortunate to have a user, Mark, who not only had expertise in the accounting area, but also was able to write excellent specifications. He translated the system functions (that had been identified in the Management Definition Guide) into the following prototype screen flow.

Fixed Assets Proposed Screen Flow

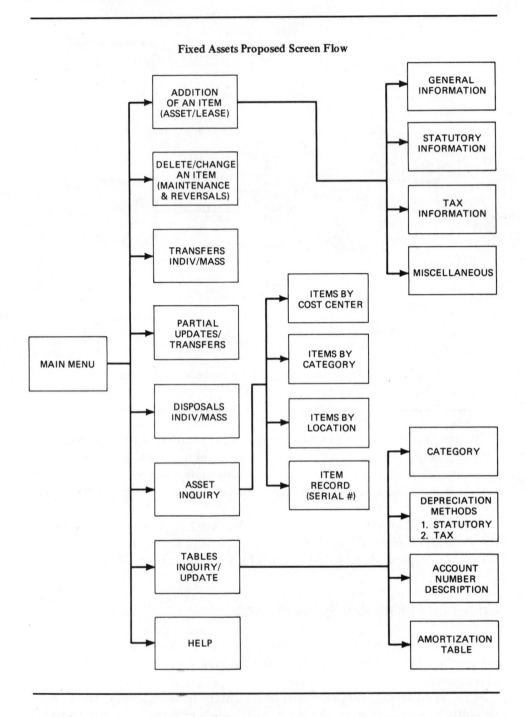

Also, he wrote screen descriptions that listed the fields each screen might contain. The following shows a portion of this list:

Screen Descriptions

Screen Name (First Level)	Screen Name (Second Level)	Screen Fields	
Addition of Item	General Information	Asset Type Company Asset Number Location Cost Center	Acquisition Date Purchase Price Voucher Number Quantity
	Statutory Information	Depreciation Period Depreciation Method	Depreciation Start Date Depreciation Status
	Tax Information	Depreciation Period Depreciation Method Depreciation Start Date	New/Used Federal Category Code
	Miscellaneous	Vendor Name Maintenance Amount Maintenance Effective Date	Maintenance Period Last Repair Date Repair Cost

For this project we had to address processing requirements. This involved gathering specifications for calculations and reversal processing considerations. For calculations, the Accounting Systems department gave us an existing document that specified all calculations required for the various groups of fixed assets. The following shows an excerpt from this list:

Depreciation Calculations

- *Additions.* Depreciation will be calculated (and the corresponding accounting entries made) as follows for all additions during the current month:

$$\text{DEPRECIATION} = \text{SYSTEM DATE (MONTH/YEAR)} - \text{PURCHASE DATE}^*$$
$$((\text{PURCHASE PRICE} - \text{SALVAGE VALUE}) / 144)$$

- *Reversal of disposals.* When a disposal is reversed via the Correct/Change screen, the following calculation should be made for adjusting depreciation:

$$\text{DEPRECIATION EXPENSE} = (\text{SYSTEM DATE} - \text{DISPOSAL DATE})^*$$
$$((\text{PURCHASE PRICE} - \text{SALVAGE VALUE}) / 144)$$

For reversal processing, we prepared a chart that identified which accounts were affected when users needed to undo (that is, reverse) transactions already entered into the system. The following shows a portion of this chart.

Possible Changes	Accounts Affected
Purchase price	Chargeback Expense
	Depreciation Expense
	Accumulated Depreciation
Purchase date	Chargeback Expense
	Depreciation Expense
Disposal price	Gain/Loss
Disposal date	Gain/Loss
	Chargeback Expense
	Depreciation Expense
	Accumulated Depreciation

For reports, we listed the existing reports and assembled the samples that the programmer/analyst had printed for us. The following shows part of this list:

REPORTS

Home Office Furniture

- Accumulated Depreciation
- Transferred to ABC—Depreciation Furniture
- Detail List by Category ID
- Transaction Listing—Home Office Furniture
- Interface Control Reports

Field Furniture

- Cost Center Report
- Items Location Report
- Categories within Cost Center

The JAD Session Agenda

Based on all the information gathered so far, we knew that the session agenda should include:

- Existing work flow
- Assumptions
- Current open issues
- New work flow
- Screens
- Processing (calculations and reversal processing)
- Reports
- New open issues
- Distribution list for the final document

The sequence varies somewhat from the standard session agenda. Instead of opening with assumptions, we began with existing work flow. We felt this allowed people to better understand the existing system before discussing the assumptions relating to the work flow.

Another departure from the standard format was that open issues were separated into two parts: *current* open issues near the top of the agenda and *new* open issues at the bottom. Before talking about the new work flow, we needed to resolve certain open issues in the Working Document. For example, one open issue asked, "Should furniture for the home office be tracked by workstation or by component part?" The outcome of this issue had a major impact on the work flow, screens, and reports.

Now, with the leaders familiar with the system, the work flow documented, the preliminary specifications gathered, and the session agenda finalized, we were ready to begin Phase 3: Preparation.

PHASE 3: PREPARATION

In this phase, we compiled the Working Document, trained the scribe, prepared the visual aids, and set up the meeting room.

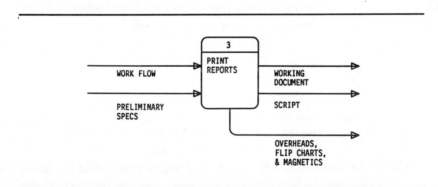

The Working Document

To prepare the Working Document, we compiled all the specifications we had gathered so far. This included existing and new work flows, screen descriptions and screen flow, existing reports, and processing requirements for calculations and reversal processing. Also, assumptions and open issues had been accumulating.

Using standard templates, we put all this information into the Working Document. We distributed it one week before the session.

The Session Script

We used the completed Working Document as the basis for determining what we would cover for each agenda item. We did not, however, create a script as described in this book. At the time, we had not yet developed the scripting technique. Instead, we used a more informal approach. The kinds of items that would normally go into a script were noted in the margins of our Working Document. This "wing it" approach worked, but not as well as using a script.

Training the Scribe

We reviewed the Working Document with Phil, the scribe, and summarized how we planned to document each part of the agenda. We showed him when to update information directly in the Working Document and when to use a particular scribe form. We asked him to come early to the session to help set up the forms. He also helped us carry the overhead projector and all the other JAD paraphernalia from the storage closet to the JAD room. He stayed each day after the session to review the scribe forms with us. And when it came to hanging the flip charts, we were glad to have a scribe who was six feet tall.

Visual Aids

We prepared flip charts for the session agenda, session objectives, and open issues. We made overhead transparencies for the data flow diagrams, screen flow, and reports. Since this JAD did not cover data elements, we did not have the usual stack of magnetics. In fact, we used no magnetics at all.

Pre-JAD Session Meeting

We did not have a full pre-JAD session meeting. This is unusual, and we thought there was a good reason for it at the time. This was the first JAD held for the financial area of the company. None of the participants had ever been in a JAD before. Because everyone was new to the methodology, we had spent time with smaller groups of users giving them the full overview of the methodology. Therefore, we did not feel a pre-JAD session meeting was necessary. Instead of distributing the Working Document at this meeting, we sent it through the mail.

What did we sacrifice by not having this meeting? First, there was the five-minute overview that would have been given by the executive sponsor to show management commitment. In this case, though, the executive sponsor was a participant in the session, so we knew that commitment would be evident. Second, we missed distributing the Working Document in person and walking the group through that document to explain what areas to concentrate on in their review. Third, the participants had no chance to establish group identity. The session was the first time they were all together. Many had not met each other before then.

Although this JAD was still successful without this meeting, we feel that if we were doing it again, we would have held the meeting.

Setting Up the Meeting Room

The challenge for the room setup was dealing with the fact that we had to hold the session in a different room each day. Instead of being able to leave our flip charts hanging and decisions written on the board over-night, we had to undo everything in the afternoon and set up again the next morning. (Here is where the extra roll of masking tape comes in handy.)

We arranged special access to the room the day before the session (which was a Sunday). Murphy's Law ("If anything can go wrong, it will go wrong") is not relaxed on weekends or holidays. The security guard could not find the letter from building management giving him the authority to let us in the room. While we waited, our carload of equipment and supplies remained precariously parked in a loading zone. Half an hour ticked by. Eventually the guard found the letter and we finally got in.

We spent a couple of hours setting up the room. We hung the flip charts, set up the overhead projector, and readjusted the table arrange-ment. We checked the pens, bulbs, and lighting for the room, and put blank name tents at every place.

At this point, we had prepared the Working Document, pseudo-script, and visual aids. We had trained the scribe and set up the meeting room. And we had decided *not* to hold the pre-JAD session meeting. We were ready for Phase 4: The JAD Session.

PHASE 4: THE JAD SESSION

For the JAD leader, the hours before a JAD session are like the countdown to an opening night of a play or the delivery of a public address. Yes, we had some butterflies. And 8:30 A.M. was only minutes away.

The group was reserved at the start. People whispered questions to each other like, "Who's the one sitting next to Mary Lynn?" They made observations about the food ("Ah, bagels . . ."). With the executive sponsor there, most people were in the presence of their boss and, in some cases, their boss's boss. Some of the people talked quietly amongst themselves ("What, no lox?") while others just filled in their name tents and waited. 8:30 arrived. We started on time.

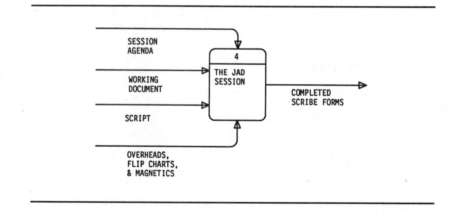

Opening the Session

We began with the administrative items. We explained where to make phone calls, when the session would begin and end each day, and when we would take breaks. We introduced the scribe and the executive sponsor.

We reviewed the agenda, explaining the importance of resolving certain open issues at the beginning of the session because the other agenda items would depend on the outcome. Finally, we reviewed the Management Definition Guide, reaffirming the purpose, scope, and objectives of the project.

Review Existing Work Flow

For existing work flow, we simply reviewed what already existed. Using the overhead projector, we presented the data flow diagrams that had been created in the smaller meetings before the session. We presented three levels of diagrams. The first level was the context diagram, which showed the Fixed Assets system with its inputs and outputs. It included all external entities, which, in this case, were the users and interfacing systems. The second level showed the three subsystems (one for each fixed assets group). There were subsystems for home office furniture, field furniture, and data processing equipment. The third level exploded each subsystem into more detail.

Someone recommended renaming some of the data flows, which we did. A few questions arose about how the system handles chargeback information. We directed these questions to the participants who were most familiar with that area of business.

Existing work flow is shown on pages 313–325 of Appendix D (the final document).

Assumptions

We came into this JAD with more assumptions than usual and by the end of the session, the list had grown longer. We reviewed each one by reading it aloud. Many of the assumptions remained unchanged while others were reworded. For example, one assumption addressed archiving fixed assets information in history files. It said,

---

To maintain a history of fixed assets information, save all transactions for a designated period of time.

Someone wanted to know exactly how long the transactions would be saved. Someone else recommended rewording the assumption to say, "Save all transactions until the tax year is closed." Several other participants suggested adding some additional archival requirements. The final assumption read:

Archival Requirements

To maintain a history of fixed assets information:

- Save all transactions until the tax year is closed.
- Microfiche all reports.
- Track all items in inventory. Delete disposed items from the data base and archive them for future reference.

Another assumption roused considerable discussion. The group could not decide how to track items that were not being used, such as spare computer terminals stored at the data center. How would these "out-of-service" items be identified and who would be charged for them? We recommended turning the assumption into an open issue to be discussed at the end of the session.

By this time, we were feeling pretty good. As they say, we were "on a roll," plowing through the agenda, making decisions while the scribe recorded every agreement. The way things were going, we figured we would finish the session a day early. We were overly optimistic, however. Soon enough, things bogged down.

Current Open Issues

The discussion on current open isues began well enough. The group resolved the first two issues. The first issue asked:

Will the system handle fixed assets information for the company's fleet of automobiles?

This was resolved to:

The system will handle fixed assets information for vehicles and buildings.

The second issue asked:

Will the Fixed Assets system calculate EDP chargeback rates? If not, will the system feed the rates to the EDP chargeback system?

This was resolved to:

EDP chargeback rates will not be calculated by the Fixed Assets system. Instead, the chargeback rate will be *manually* loaded into the system. Then the actual monthly chargeback amount for each cost center will be automatically fed to the Chargeback (AER) system. The chargeback rate will be included in the Category Table.

The third issue asked:

Should the Fixed Assets system also track inventory for all or certain classes of assets?

After discussing this issue for 10 minutes, we asked if it needed to be resolved before going on to the next agenda item. The group agreed it did not, so we left it as an open issue to be discussed at the end of the session. We noted it on the flip chart. Then we went on to the fourth issue—and soon were reminded of the main reason for moving issues to the front of the agenda. This was the question that would besiege us for the rest of the morning. The fourth issue asked:

Should furniture for the home office be tracked by workstation or by component part?

After the group discussed this question for 15 minutes without resolving it, we asked, "Do we need to resolve this now?" Someone in the group insisted, "Yes, the whole design of the Fixed Assets system depends on how this issue is resolved. It affects every remaining item on the agenda." The group unanimously agreed. We had no choice but to continue wrestling with the issue, then and there.

Twenty minutes later, we realized we were getting no closer to a resolution. It was time to resort to the formal techniques for resolving conflicts. We had already opened up the question to the group. That was an understatement; everyone was wholeheartedly involved. We had already suggested making it an open issue, but that was out of the question. We didn't have to call the executive sponsor; she was already in the room and was just as interested in resolving it as the rest of us. So, we did the only sensible thing—we called a break.

Unlike other JADs, where major issues were resolved (or at least clarified) during small break-time discussions, everyone returned from the coffee and bagels just as perplexed as before.

Analyzing the nature of the discussion, we saw that the issue was not so much a *conflict* as it was *indecision*. In other words, it was not that one group of people was strongly advocating one particular side of the issue. Everyone just wanted to determine the best approach. The problem was that there were so many variables—so many advantages and disadvantages to each side that it was difficult to see the whole picture.

Now was the time to take pen in hand and use the technique described in the "How to Handle Indecision" section of Chapter 8. For this issue, our alternatives were: (1) track furniture by workstation; or (2) track it by component part. We determined the evaluation would be based on the following criteria:

- Accuracy of the subsidiary ledger
- Work versus return on investment
- Chargeback accuracy
- Cost control
- Ease of system construction
- Ease of system conversion

Then we determined weight factors and scores. We calculated the weighted score and came to a resolution. Tracking inventory by component part was the winner by a moderate margin. The chart on page 219 shows the resulting assumption that became part of the final document.

By now, the morning had turned to afternoon and the day's session was coming to a close. Although we had made it through the biggest hurdle of the project (resolving how home office furniture would be tracked), we had not accomplished as much of the agenda as we planned. The first of four days was gone, but we were only part way through the third agenda item (current open issues). The only consolation was knowing from expe-

HOME OFFICE FURNITURE INVENTORY METHODOLOGIES

The participants evaluated the following inventory methods to use for Home Office Furniture:

- Method A (Inventory by Workstation). This is the method currently used when components are tracked by workstations; for example, "A" stations, "B" stations, and so on.
- Method B (Inventory by Component Part). For this method:
  - Inventory will be tracked by component part at the company level.
  - Chargeback will be done by the Accounting, Budget and Cost (ABC) system.
  - Each responsibility center will have a chargeback amount based on a more general formula (for example, square footage or staffing) as opposed to keeping a specific inventory by cost center.

The following chart shows the participants' evaluation of methods A and B:

- The first column *Criteria*, shows the criteria identified by the team for evaluating the two methods.
- The second column, *Weight Factor*, shows the relative importance given to each criterion. The weights are 1 through 3, where 3 is the most important.
- The last two columns show the two alternatives:
  - The *Score* columns show the relative rating, where 4 is the best. For example, the criterion "Ease of system construction" rated fairly high (3) for Method A and low (1) for Method B.
  - The *Score × Weight* columns show the numbers resulting from multiplying the Score by the Weight Factor. This gives a relative weighted rating where a higher number means a better rating.

Criteria	Weight Factor	Method A Fixed Assets by Workstation		Method B Fixed Assets by Component Part	
		Score	Score × Weight	Score	Score × Weight
Accuracy of the subsidiary ledger	3	1	3	3	9
Work vs. return on investment	2	1	2	3	6
Chargeback accuracy	2	3	6	1	2
Cost control	1	3	3	2	2
Ease of system construction	1	3	3	1	1
Ease of system conversion	1	3	3	4	4
Totals			20		24

Based on this evaluation, Method B is the best approach.

rience that this is typical of the first day of a session. We were familiar with
the oh-my-gosh-we'll-never-finish-the-agenda syndrome that often occurs
on day one of the session. Dealing with major issues that affect the rest of
the agenda takes time. Once they are resolved, however, the rest of the
session moves more swiftly.

On day two, we discussed the remaining open issues in only 45 minutes.

New Work Flow

For new work flow, we used the overhead projector. We presented the
transparencies of the data flow diagrams we had created in the meetings
before the session. We explained how these diagrams were produced and
who was involved. We emphasized that this work flow was proposed and
that the group would review and change whatever was necessary.

We explained that we would go through the diagrams twice: once for
review (to make sure everyone understood them) and another time for
update (to make changes). In the first pass (the review), people had
several questions regarding which departments source documents would
come from. In the second pass (the update), we spent more time on each
diagram. Minor changes to the work flow were considered and agreed
upon.

Screens

As we said earlier, this RFP JAD did not require detailed screen design.
Therefore, this part of the agenda was straightforward. Mark, our "user
analyst," had already prepared the screen flow diagram and screen descriptions that were included in the Working Document. Consequently,
the "screens" review was limited to discussing the screen information and
getting formal agreement. Usually, JAD sessions involve some kind of
screen design. For a description of a typical episode, see the "Screen
Design from Scratch" section of Chapter 12.

Screen flow and descriptions are shown on pages 327–329 of Appendix D.

Processing

For this JAD, processing requirements included calculations and reversal
processing.

We did not review the calculations one by one. Instead, we asked, "Does
anyone have any questions on the calculations for depreciation?" Pause.

"How about for gain/loss?" Pause. And so on. When we reviewed the monthly chargeback calculations, someone suggested replacing the words "monthly maintenance amount" with "expensed item chargeback rate." A few of these minor changes were all that were required for calculations.

For reversals, we first reviewed all the changes that could possibly arise from having to reverse a transaction. Then, for each reversal, we identified which accounts would be affected. All this information was added to the Working Document (in the chart with the two columns *Possible Changes* and *Accounts Affected*).

Processing requirements are shown on pages 327 and 330–331 of Appendix D.

Reports

The Working Document contained samples of all existing reports. We reviewed each one, first to make sure everyone understood its purpose, then to decide whether to keep it, change it, or get rid of it. We did this by paging through the Working Document.

Then we identified new reports. As the list grew, we analyzed which reports were truly required as opposed to just "nice to have." In discussing what general information these reports would contain, we identified some requirements that could be handled by adding a new column or two to already existing reports.

For most JADs, we would have gone on to design these new reports. But for this RFP JAD, we needed only to identify them. The list of existing and new reports is shown on pages 331–335 of Appendix D.

New Open Issues

At the end of the session, we had seven issues listed on the flip chart. One was from an assumption that had turned into an open issue. Another was from the list of current open issues that we covered on the first day and could not resolve. The other five came up during the session. We tackled these issues one by one. This was the outcome:

- Through discussions, five issues were resolved and agreements were made. Again, the scribe documented the resolutions on Assumption scribe forms.

- The remaining two issues could still not be resolved. In fact, the group agreed that they required more research and would be better handled in separate meetings after the session. They were assigned to the appropriate participants to resolve by a certain date.

These two remaining issues are shown on pages 334 and 336–337 of Appendix D.

Distribution List for the Final Document

We asked who else besides the participants needed copies of the final document. Four names were added to the distribution list.

Closing the Session

To close, we had only to summarize what was to happen after the session. We explained that we would be creating the final document from the scribe forms and the changes noted in the Working Document. We would distribute a draft copy to all the participants for their review. Then, we would reconvene to discuss any changes. If the changes were major, we would reissue the document.

Ordinarily, the JAD wrap-up would end here. But, since the contents of this document would be used to build the Request for Proposal to send out to the vendors, we thought it would be worthwhile to discuss the steps of the RFP process. So we guided the participants through defining and scheduling the following steps:

1. Users prepare the RFP.
2. Vendors submit their proposals.
3. ISD submits its proposal.
4. Users select vendor or in-house solution.
5. Users either purchase a package (for the vendor solution) or hold another JAD session (for the in-house solution).

The session came to a close. The 18 participants, who began the session feeling unsure about spending four half-days doing this thing called JAD, now appeared to feel good about what was accomplished. (Or were they just elated to finally be free of the hardback chairs they had been sitting in?) Whichever it was, they went back to their "life before JAD" and we went on to prepare the final document.

PHASE 5: THE FINAL DOCUMENT

In this phase, we produced the final document, held a full-team review meeting, and obtained final approval on the specifications.

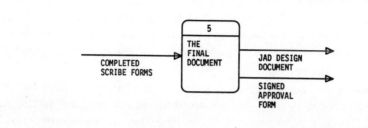

Producing the Final Document

Producing the final JAD Design Document involved updating the Working Document. To do this, we pulled the information from the following three sources:

- notes from the scribe forms filled in during the session
- notes the scribe made in his copy of the Working Document
- data flow diagrams. Using our CASE tool, we updated the data flow diagrams to reflect the changes in the work flow.

We prepared a final version of the document. This took three days. Then we ordered enough copies for all the participants, put them in the binders, and added the preprinted tabs for work flow, screens, reports, and so on. We inserted the cover page, back page, and the preprinted spine tabs that said "Fixed Assets." One week after the session, we distributed these documents along with a cover memo that confirmed the time and place for the review meeting. Appendix D shows the final document.

Straying from the Guidelines

In preparing this final document, we strayed from our JAD guidelines and wished we hadn't. Here is what happened.

This particular blunder related to open issues. While we were preparing the final document, a participant called and said that he had resolved his open issue, "What are the Information Services Operations requirements for the Fixed Assets system?" He had documented the resolution based on a meeting held with the others assigned to that issue. He asked if we would include the outcome in the final document. We had never done this before. But in an effort to be flexible, we said, "Certainly, just send along the resolution and we will put it in the document."

We decided to call another person who was handling the other open issue that was scheduled to be resolved that week. This issue was, "Will the Fixed Assets system support both capital and operating assets?" We told him that if he resolved it by the time we sent out the document, we would be glad to include the resolution.

What happened is this: The first person sent his resolution to us the next day. The second person (whom we had called) worked particularly hard to finalize his resolution. We, in the meantime, completed the document ahead of schedule. Since we did not hear from the second person, we assumed his resolution was not coming. Then, just when we completed the document and sent it off to the mail room, he called to say his issue was resolved and he would send it right away. We told him the documents were already upstairs in the mail room ready to go out. Naturally he was annoyed for having made the extra effort to resolve that issue ahead of time. So, up to the mail room we went to pull the documents, undo the packaging, insert the new resolution, repackage them, and send them out a half day later.

Our efforts to be flexible were more a burden than a help. The final document should represent a point in time—the end of the session. The document should not include decisions made afterwards. Such is the challenge of keeping the balance between staying flexible (making everyone happy) and following the JAD guidelines.

The Review Meeting

Everyone attended the review meeting and almost everyone had a change or two to offer. The group agreed that sending out an updated copy was not necessary. We decided, however, to send out some of the revised pages where charts were changed. Normally, we would not have done this, but since the users planned to include these charts in the Request for Proposal that would go out to the vendors, we updated the charts and distributed them in a final mailing.

Approving the Document

This was the simplest step in the whole project. Everyone agreed with the contents of the document as well as the changes made during the review meeting. The participants designated to sign the Approval form did so without arm twisting, threats, or other forms of duress. We sent out copies of the signed form to all participants and kept the original with the Assistant VP, Applications Development.

THE OUTCOME

So, how did it all turn out? Was inventory tracked by component part (as determined in the session) or did the other method (inventory by work-station) resurface as an option? As the session decided, the component part method was used for all inventory from that time forward. And did we go with the PC package or the mainframe solution? The company purchased a PC package. The JAD Design Document was used extensively in the package evaluation.

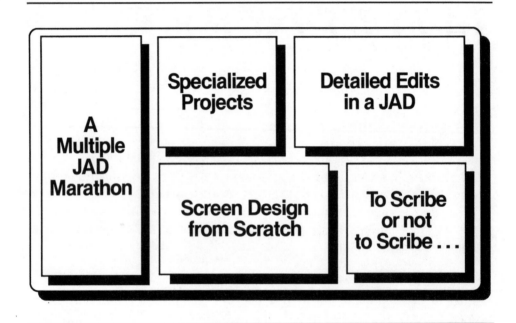

A
Multiple
JAD
Marathon

Specialized
Projects

Detailed Edits
in a JAD

Screen Design
from Scratch

To Scribe
or not
to Scribe . . .

TWELVE

JAD EPISODES

The Fixed Assets project described in the case study illustrates many aspects of the JAD methodology, but by no means all of them. To show you some additional highlights (and low points) of our experiences, this chapter describes several episodes from our 30 JAD projects. Among them is a multiple JAD that we did in a time crunch, a project that addressed nothing but data elements, another project that handled only transactions, a project where we designed screens from scratch, and a political situation we encountered regarding who should be the scribe.

A MULTIPLE JAD MARATHON

Of all the projects we have done, no other was larger or had a time constraint as tight as this one.

The project involved defining specifications for a new insurance product that allowed policyholders to custom tailor an insurance plan by combining the features of whole life and term insurance in one policy. This product also allowed policyholders to change or add to their existing coverage (without having to buy a separate policy) as their needs changed over time. To support such a flexible product, complex specifications needed to be defined for five separate computer systems.

To meet the time crunch, we ran 25 back-to-back sessions—not half-day, but *full-day* sessions. That's a lot of Danish! We were running sessions for one JAD, preparing for the next JAD, and producing the document for the JAD before, all at the same time.

The Time Constraint

In December, the Business Systems Engineering group of our company was asked to become involved in this project. To meet the deadlines, the programming staff needed all the JAD Design Documents by March of the following year. Thus, we had four months to do the JAD work. Somehow, between Christmas and the Ides of March, we had to become familiar with the new insurance product and its effect on five computer systems, prepare for the sessions, conduct the sessions, and produce the JAD Design Documents. Ultimately this involved making 25 flip charts, 90 overheads, 150 magnetics, completing hundreds of scribe forms, and generating 748 pages that made up six final documents.

The People

Our resources included two JAD leaders and one-and-a-half support people from BSE (one was working on other projects as well, so she was available part time). Also, we had the help of an administrative assistant for such tasks as copying handouts, preparing overheads, and assembling the final documents.

Approach

We divided the project into six JADs, one for each of the five application systems and one to define the data elements for the data base. Each of the six sessions lasted two to four full days, and each had its own list of participants. We could not run any sessions concurrently because some participants were directly affected by more than one system. Therefore, they needed to attend several JADs. In fact, some people attended every session of every JAD.

For each separate JAD, we assigned a leader and a support person to follow through from the first interviews to the creation of the final document. Also, both leaders participated to some degree in each JAD project. This "cross-familiarization" was necessary so that one leader could fill in for another, if necessary. In fact, during one JAD session, the leader could not make it into work because of a major snow storm. The second leader filled in.

To handle this substantial project, we made adjustments to our standard JAD practices. The following describes them.

Phase 1: Project Definition

- *Interviewing management.* The challenge was scheduling. The people we wanted to meet with for the upcoming JAD were often attending the current JAD. We arranged evening and weekend interviews.

- *Selecting the JAD team.* We developed one master participation list that we updated regularly. It included a running total of participants for each JAD as well as a grand total. We used this list to see at a glance who was attending what sessions, to determine room size and seating, and for distribution of memos and the final document.

- *Management Definition Guide.* This project had such priority that a high-level "Working Committee," run by the users, met regularly to plan and track its progress. They maintained a project plan and determined the purpose, scope, and objectives of the project. There-fore, we did not prepare a Management Definition Guide.

- *Scheduling the JAD.* We kept one master schedule to track meeting room locations. We held all six sessions in one off-site facility, although many times we had to change rooms from one day to the next.

Phase 2: Research

- *Familiarization interviews.* We shortened the time for observing the existing systems in action. Such familiarization was a luxury that the time constraint did not allow. Instead, we focused on understanding the preliminary product specifications that the actuaries and the ISD project managers had prepared for the new insurance product.

- *Documenting work flow.* We created some overview work flow and analyzed the work performed in the user area. But we did not do work flow to the detailed level as in other JADs.

- *Gathering preliminary specifications.* We handled this as usual, gather-ing as much as we could on data elements, screens, reports, and processing requirements.

- *Preparing the session agenda.* As always, we prepared the agenda before each session. Normally this is done at least two weeks before the session. For this project, however, the agenda was sometimes not finalized until the day before (or sometimes minutes before) the session.

Phase 3: Preparation

- *The Working Document.* Here is where we completely changed pro-cedures. Since the BSE staff did not have time to prepare an official

Working Document before each of the six sessions, we relied on the project managers to compile preliminary specifications. These were usually prepared on a PC and were not brought together into the document format until the final JAD Design Document.

- *Preparing the script for the session.* No formal scripts were prepared. We had not, at that point, developed the scripting technique.

- *Training the scribe.* For several of the six JADs, the scribes themselves entered much of the scribe form information into the files for the final documents. This was a great help, but it meant the JAD leaders had to spend more time training each scribe. We needed to cover not only the scribing process, but also how to use the standard templates and the text editor.

- *Visual aids.* We never skimp on these.

- *The pre-JAD session meeting.* These were not held.

- *Setting up the meeting room.* Since the sessions for all six JADs ran back-to-back, we transferred the supplies to the facility in one move and stored them in a locked closet throughout the 25 days. Our presence at the facility became so consistent that we were soon "promoted" to residency status (which means we received our own keys to the storage closet and restrooms).

Phase 4: The JAD Session

The main difference in the sessions was that they were full days instead of half. The support person working on a particular JAD attended the session in the mornings (to follow what was going on) and worked on preparing the final documents in the afternoons. Also, the support person spent time with the scribes who, as described earlier, played a stronger role than usual in preparing the final documents.

Phase 5: The Final Document

- *Producing the final document.* The challenge here was contending with the snowball effect. After the first JAD, we produced the document within days. After the second JAD, we spent a little longer completing it because we also had to prepare for the next JAD. Since preparing for the session was the number one priority (you have to be ready to walk on stage), the final documents could easily have been slighted or put on the back burner. To avoid this trap, we worked overtime to keep up with the documents.

- *Assembling the final document.* Rather than send separate binders for each JAD document, we combined the documents into one larger binder, using tabs to separate each project. People attending four of the six JADs, for example, received a 2-inch binder containing the documentation for all four projects.

- *Tracking distribution.* Tracking distribution was more critical for this project than any other. One participant might receive one final document, while another might receive all six. Since documents were being sent out every week or so, we had to track who already had binders, what size binders they had (1-inch, 1-1/2-inch, or 2-inch), and how many documents they received. This major tracking effort would have been impossible without documenting who received what and when.

- *The review meeting.* Because the sessions for one JAD began as soon as another one ended, we could not hold review meetings. Some of the people who would have attended those reviews were participating in the next session, so we could not have scheduled everyone together. Instead, we selected two or three key participants from each JAD to review the document for accuracy prior to distribution.

- *Approving the document.* Since not everyone in each session was able to review the final document, we felt it was not appropriate to ask for these signatures. Instead, approval was in the form of verbal acceptance by the key participants who reviewed the documents.

The Wilt Chamberlain Syndrome

Perhaps the biggest challenge of doing a project with tight time constraints occurs long after the project is done. If you succeed in pulling it off, some people assume that you can work at that pace on an ongoing basis. Therefore, they might expect the same from you again.

In a basketball game in 1962 between the Philadelphia Warriors (now the 76ers) and the New York Knicks, Wilt Chamberlain scored 100 points. His fans, of course, wanted to see this extraordinary feat again. They now figured that when Wilt had a 50-point game, he was doing only half of what he could. Even if they forgot about the 100-point game, there was still his average to consider: 50 points per game for a whole year. As he said in his book, *Wilt,* "That meant any time I hit 30 points in a game, it was a 'bad' night: I'd have to get 70 the next night just to make up for it and stay even."

Although you can't quite compare a JAD session to a basketball game with "Wilt the Stilt," the resulting syndrome is the same. Try not to be pressured by other people's expectations that were set during a successful but overly intensive round of JADs.

In Closing

So that is the saga of our six-project, 25-day-session JAD marathon. The JAD methodology was essential to completing this project on time. As the department Vice President said,

> Our new comprehensive life product could not have been produced on time without the JAD process because of the difficulty of the specifications and the tight schedule established by the customer to have the product in the field.
>
> —Guy Edwards
> V.P., Information Services Department
> Provident Mutual Life Insurance Company

Of course, we recommend spreading out the schedule quite a bit more. Even adding one week between sessions would have made a world of difference. Nevertheless, we have summarized this crunch-mode project for the sake of those who may fall into such a situation when upper management says, "We need this project done by January 1."

SPECIALIZED PROJECTS

Once your organization has had a taste of the JAD methodology, you might be commissioned to hold JADs for atypical, more specialized projects. As long as the requested projects meet the JAD criteria that you have set up, be open to breaking away from traditional JADs. The following describes two such projects, one for defining data elements, another for defining transactions.

One JAD, Another Hundred Data Elements

Why even mention this project, which contained only one main item from a normal session agenda? We mention it to show how specialized JADs can really be.

This project was the first of the six-JAD series described in the previous section. The session had one objective—to define all the data requirements for the new insurance product. In other words, what should the data base contain to support the systems' enhancements? We also handled some data base design, assumptions, and open issues; but the main focus was on data elements.

By the end of the four-day session, the group had added five new records and 45 new data elements to the policy administration data base and modified 15 existing elements. For each data element, we identified the name, format (alphabetic, numeric, or alphanumeric), and length. We provided definitions that ranged from half a page to two full pages. All this information went into the final document, which was an essential prerequisite for several of the remaining JADs in the series.

In this JAD, we developed five customized scribe forms to handle record definitions. The forms identified:

- which data elements would be contained in each record type
- expected record volumes
- the overall logical data base design

Samples of these forms are shown in Appendix B. Normally, data base design is handled by the data base and programming staffs after the JAD. But to fully understand the impact of the new insurance product on the critical systems, we had to perform some data base analysis in the session.

A JAD of Transactions

The company wanted to supply its clients with a monthly consolidated statement for all products each client has with the company. This would allow clients to review at a glance a summary of the current status of their policy information and the significant transactions that were processed in the reporting period. This JAD dealt with selecting and interpreting all the existing transactions from the policy administration system that would feed this new client system. The transaction type would dictate the wording to print on the client statement. We prepared a customized scribe form. A filled-in sample of this form is shown in Figure 12-1. We selected a scribe from the user area who was an expert on the various policy administration transactions.

The session was certainly not the most inspirational endeavor. Can you imagine going through 80 separate transactions, filling out the same scribe form for each one? Nevertheless, the users felt it was worthwhile, and the programmers got all the information they needed to program the system.

This was one of several projects where the programmers wanted to update the final document as the specifications changed over time. Transactions were constantly being added as clients requested new system features. We resolved this by giving copies of all our text files to the

---

TRANSACTION DESCRIPTION FORM

Transaction _____1_____

Transaction code _____2002_____

Transaction type _____HISTORICAL_____

System source _____CFO_____

How generated _____EXTERNAL_____

Description _____PREMIUM PAYMENT_____

Other parameters _____AGENCY 13_____

Variables in the wording   _____V1 - TIME PERIOD (1-12 MONTHS)_____
_____V2 - PREMIUM DUE DATE_____

Wording on the client statement   _____PREMIUM FOR (V1) MONTHS DUE_____
_____BY (V2). THIS AMOUNT IS PAID_____
_____BY THE COMPANY._____

Amount description   _____GROSS MODAL PREMIUM_____

Figure 12-1 Scribe form for the Transactions JAD

programmers, who updated them with each new enhancement. The original files stayed with us, unchanged.

DETAILED EDITS IN A JAD?

Depending on the nature of the JAD, you may be asked (usually by MIS) to do some internal design. Usually these details can be handled outside the session in smaller meetings with the programmer and a couple of users. There are times, however, when it is helpful to have both users and MIS together in a room to define processing, or what happens to the data after it enters the system.

The following shows one example where we delved into calculations and another where we defined some edits.

Calculations

We have defined calculations in several of our JADs. One such project involved the users and MIS people defining how various insurance pieces

Death Benefits

When calculating death benefits:

- For death within 31 days following any unpaid premium, charge the pro rata part of the full modal premium up to the death date.
- Allow scheduled increases up to the death date and charge the appropriate premium to the death date.

MODAL PREMIUMS

Given a due date and mode, the calculation determines a modal premium based on:

- rate data
- historical event data
- scheduled increase data

BONUS DIVIDENDS

Inside and outside additions reserves are combined for bonus dividend calculations. When dump-ins occur during the policy year, use weighted gross funds based on the number of whole months between the effective date of the dump-in and the anniversary. When dump-ins occur after the dividend has been calculated, pay no bonus dividend on the dump-in for that policy year.

Figure 12-2 General calculation definitions

(such as premiums and dividends) would be calculated for the new product. These general calculations did not include actual formulas. Figure 12-2 shows a sample from the Policy Administration JAD. Although the terminology may not be entirely clear (it is insurance-specific), you can get an idea of the kind of information that was captured.

In another project (the Fixed Assets case study), we defined calculations in such detail that they were documented as formulas. We did not bog down the session by defining these formulas in the presence of all 18 participants. Instead, the formulas were prepared beforehand and reviewed for accuracy and completeness in the session. Figure 12-3 shows a portion of these calculations.

Edits

Edits are checks performed by the system on data entered by users via online screens or batch transactions. Edits verify whether the entered data conforms to the specifications set up for that data.

One JAD project involved enhancing an underwriting system to sup-

Depreciation Calculation

Depreciation = system date (month/year) LESS purchase
      date (month/year) TIMES ((purchase
      price LESS salvage value) DIVIDED BY
      depreciation period))

MONTHLY DEPRECIATION CALCULATION

Depreciation (monthly) = (purchase price LESS salvage value)
      DIVIDED BY depreciation period

Figure 12-3 Detailed calculation definitions

port both individual and variable life products. This required expanding existing data element definitions to accept variable life data and creating new data elements. To document this, we built a chart listing the data elements and their corresponding values. Figure 12-4 shows a portion of this chart.

Another project involved defining the edits required for another new insurance product. These edits dealt with the relationships between the fields and their consistency with insurance product rules and regulations. Warning messages would display for the underwriters on the new business screens when certain conditions for policy approval were not met. For example, if a user entered a face amount for a new policy that exceeded a certain limit and no chest x-ray was included with the application, the system would display a warning: "Chest x-ray needed for this amount." Figure 12-5 shows a sample of these messages.

As we mentioned, processing requirements such as these are best handled in separate, smaller meetings between the users and MIS. But if you need to address them with all the participants, gather as much informa-

Data Element	*Values*
Dividend Option	C = pay in cash P = reduce premium U = unscheduled premium
Dump-in Separate Acct	1 thru 8

Figure 12-4 Data feed edits

Edit No.	Warning Message
0050	Aviation questionnaire form needed
0051	Hazardous sports questionnaire needed
0052	HO specimen needed for this amount
0053	Electrocardiogram needed for this amount
0054	Chest x-ray needed for this amount
0055	Need blood profile, urinalysis, cocaine screen

Figure 12-5 Data relationship edits

tion as possible before the session. Then that agenda item will become a matter of review rather than one that drags the participants through identifying a long list of processing requirements.

SCREEN DESIGN FROM SCRATCH

In the case study, we did not cover detailed screen design because the Fixed Assets JAD went only as far as defining screen *flow*. Therefore, we will describe a segment from one of many projects where we defined detailed screen design.

This JAD project involved enhancing an underwriting and insurance policy issue system to support both individual and variable life products. It required designing new screens. We were fortunate to have the full-time help of the Variable Life project manager from Applications Development. Her analysis before the session included preparing two prototype screens.

During the session, after identifying screen flow, we displayed the overhead transparencies of the two prototype screens. We asked, "Can these prototypes be used as a starting point?" As we reviewed them, it became clear to everyone that so many data elements had been added or changed since the Research phase that most of the screen fields were now irrelevant. So, we threw away the transparencies and began with a blank board.

To design the first screen, we drew a large box on the board indicating the frame of the screen. We asked the group to title the screen and add the headers and footers. Figure 12-6 shows what the board looked like. On the left panel of the board were several columns of magnetics for all the data elements we had defined so far.

Since these screens would link into existing menus, we did not need to design new menus. We moved directly into designing the data entry

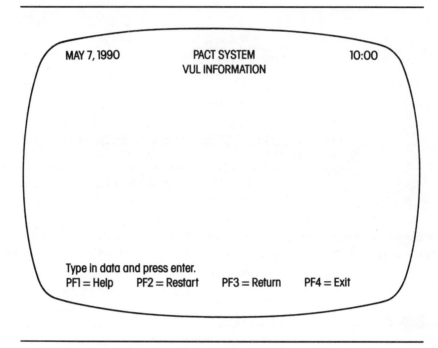

MAY 7, 1990   PACT SYSTEM   10:00
                                VUL INFORMATION

Type in data and press enter.
PF1 = Help   PF2 = Restart   PF3 = Return   PF4 = Exit

Figure 12-6 Screen design with title, headers, and footers

screens that would be built from the displayed data elements. To determine the contents of the screen (that is, the fields), we asked the group
to call out the names of the data elements they needed on the VUL
Information screen. As they did so, we moved the magnetics inside the
screen frame in no particular order.

With all the fields identified, we asked the participants to arrange
them as they should display on the screen. First, we sorted them loosely
into three groups, as shown in Figure 12-7.

Then, we fine-tuned each group. We shifted the magnetics until everyone was satisfied with the layout. We reminded them to leave space for the
field labels that would precede the fields. Figure 12-8 shows what the
board looked like then.

To define the field labels, we asked the group, "What words will
precede the fields?" When someone wanted to use the label "Sep Acct
Pct," we encouraged them not to abbreviate. There was plenty of room on
the screen to use meaningful spelled-out names. Because some field labels
were longer or shorter than expected, we shifted the field locations just a
bit. Figure 12-9 shows what the board now looked like. (The fields, or data
elements, are shown in all uppercase; the field labels are shown in upper-
and lowercase.)

With the first screen complete, we went on to design the second. We
used a separate board panel so that we could refer to the first screen when
designing the second.

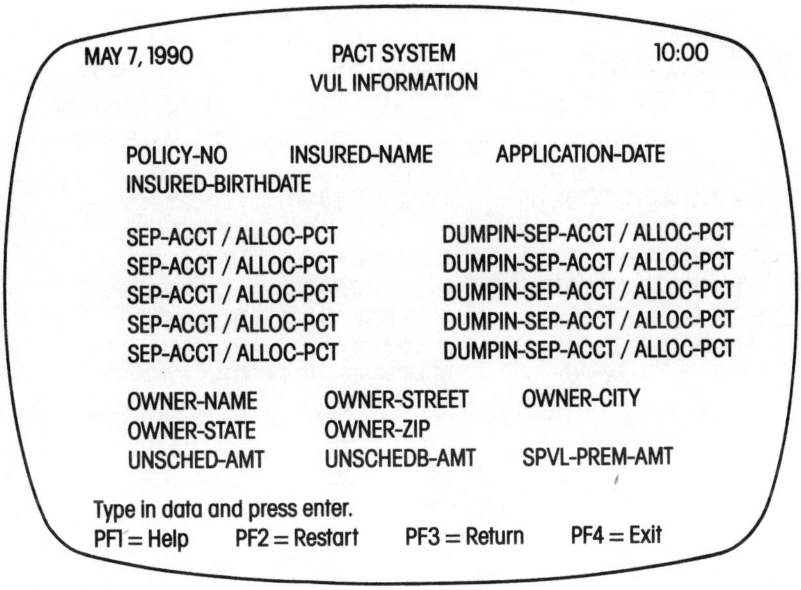

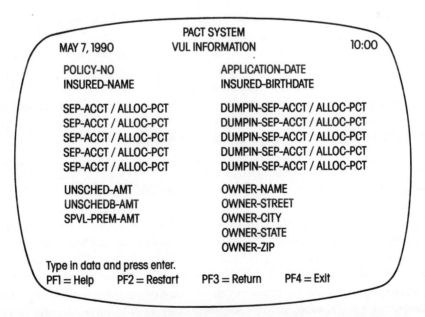

Figure 12-7 Screen design with loosely sorted data elements

Figure 12-8 Screen design with fine-tuned data elements

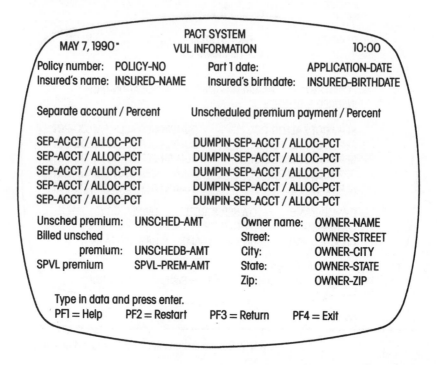

Figure 12-9 Completed new screen design

As we reviewed the completed screens, the group realized that the title on one of the screens could be more meaningful. After a bit of discussion, and some indecisive wavering between two possible titles, one good-natured, agreeable participant said, "Oh, just call it whatever you like." But we knew that we should only make suggestions; the decision (even one as simple as this) must be made by the users. After we all discussed how the screen would actually fit into the work flow, someone came up with a meaningful title and the screen design was complete.

TO SCRIBE OR NOT TO SCRIBE—A POLITICAL DRAMA

JADs are used in business, and business is never without political struggles. One that recurs from time to time in our organization relates to the scribe role.

We have mentioned how some people are not at all thrilled about taking on this role. They feel it is "not in their job description" to record decisions in a JAD session. They view the task as taking notes, and believe they left all that behind in school.

A controversy arose during one of our multiple-JAD projects. We have always selected our scribes specifically for each project. The programmer/analyst whom we selected for scribing in this case felt this was not an appropriate role. The programmer/analyst's boss (a project manager in Applications Development) defended the person by saying, "Scribing should be done by the people running the JADs, not by the programmer/analysts." As the controversy continued, several other project managers got involved, voicing their opinions on whether their people should scribe or not.

At this point, let's discuss the two scribing approaches. The question is this: Should scribing be done by someone selected specifically for the project based on his or her knowledge of the application (the temporary scribe), or by someone from the JAD staff (the ongoing scribe)? There are advantages and disadvantages to both approaches. The following describes each one.

The Temporary Scribe

When scribes are selected by project, you have people who are familiar with the application and can therefore follow the subtleties of the discussion. This familiarity makes their recording task more effective.

Another advantage is the positive by-product that comes from scribing. Because they must follow the discussion closely and word the decisions clearly, scribes naturally leave the session with a deeper understanding of the new system design. Programmer/analyst scribes find this benefits their work whether they are coding or further defining specifications. User scribes gain by having a head start on how the system works and the logic behind the design.

The disadvantage of this approach is that each time new scribes are selected, you must take the time to familiarize them with the scribe forms and the recording process. Also, when you select the scribes, you might not know enough about their abilities to be sure they can do the job well. At some point, you might end up with a scribe who knows the application but does not have the communications skills to record decisions effectively. Furthermore, even with competent scribes, you must deal with variations from one scribe to the next. Although the structure of the scribe forms dictates how information is documented, scribing styles can vary somewhat within that structure.

The Ongoing Scribe

The other alternative is to designate someone in your JAD group to be the ongoing scribe who records the decisions for every session. This way, you train the scribe only once. In fact, with each project, the scribe perfects the task, picking up tricks of the trade along the way.

Also, with an ongoing scribe, you are dealing with a known entity. There are no surprises about how that person performs. The style is consistent. This simplifies your task of reviewing the completed scribe forms after each session. Furthermore, the communications gap between the scribe and the person entering the information for the final document is completely eliminated because both tasks are done by the same person.

Now for the disadvantages of an ongoing scribe. Ongoing scribes are not familiar with the applications. Certainly their background gives them expertise in some areas, but they do not bring the in-depth application knowledge to each session that specially selected temporary scribes would. This affects their ability to document decisions in the session. Then, once the project is over, the familiarity they have gained from scribing the session has no further value. The application knowledge cannot be applied in later phases of the project, because the scribes have moved on to scribing other sessions for other applications they also know little about.

Another disadvantage relates to staffing. You must keep someone in your group to do the scribing. Since the task is intermittent, you must have other responsibilities for that person. This may not be too much of a problem. Because the communications skills required for scribing are generic, they can be applied to other tasks as well. How much of a problem it is depends on what your scribe's other duties are, and how often he or she can be pulled away for a JAD session.

The Controversy Subsides

We were able to get through this controversy with a workable solution— we used a temporary scribe. As it turned out, we did not use the programmer/analyst originally selected. Instead, we chose another programmer/analyst who would be doing most of the applications development work for that part of the project. Our scribes today are still selected specifically for each JAD. To temper the disadvantages, we try to select the right people for each application and, if they work out, we make sure we select them again when a JAD in that application area arises. Also, we spend time to make sure they understand the scribing process.

The question of who should scribe (temporary or ongoing scribes) comes down to this: Which has more importance—the scribes' familiarity with the application, or their familiarity with the process of scribing? When the controversy arose in our organization, we reexamined this question but still came up with the same conclusion we had before:

Knowledge of the application has more value than knowledge of the scribing process. We would rather show people how to scribe (a straightforward task accomplished in one meeting) than to teach them the application (an impossible task). No matter what efforts we would make to familiarize scribes with an application, they could never be as proficient as those who have worked with the system as an ongoing part of their jobs.

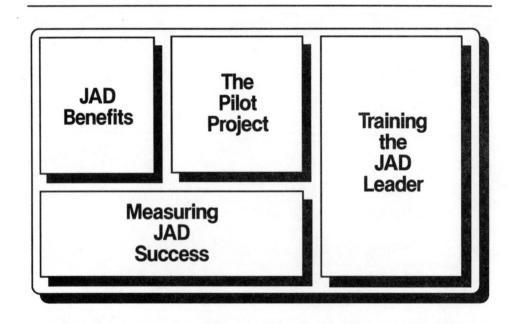

T H I R T E E N ▷▶▶▶

WHERE TO GO FROM HERE

You now have the information you need to run a JAD project. But what do you do to actually bring JAD into your organization? You can't just pick up a felt-tip pen and start making flip charts. You need to communicate the benefits of JAD to your management, select a pilot project, train other JAD leaders who will assist you, and determine ways to measure the success of the JAD methodology.

JAD BENEFITS

The benefits of JAD were highlighted in Chapter 1. This section brings them all together and offers some comments from other companies using JAD. This information can help in your own efforts to gain management support in bringing the methodology into your organization.

Many companies using JAD have recorded productivity gains from 20 to 60 percent in the analysis and design phases of the life cycle. These companies include American Airlines, IBM, Mutual of New York, CNA Insurance, Carrier Corporation, and Bell Canada. (Rush, 1985)

CNA Insurance of Chicago began using JAD in the early 1980s. A joint productivity study conducted from December 1982 to March 1983 with IBM showed the JAD methodology had boosted productivity at CNA by more than 50 percent. (Gill, 1987)

These companies, as well as our own, have found that the JAD methodology can do the following:

* accelerate systems design
* increase the quality of the final product
* improve user relations

The following describes these benefits.

Accelerate Systems Design

When faced with the challenge of defining a system with too many considerations and too little time to accomplish the task, you can turn to JAD. JAD shortens the systems development time. It reduces months of meetings to one workshop attended by everyone involved in the project. Agreements are finalized because everyone affected is there. This group consensus short-circuits the traditional approval process, where draft documents sit on reviewers' desks for days or weeks. Also, by defining more about the system in the beginning, you spend less time later having to gather additional specifications or clarifications from users to fix what was not coded correctly in the first place. The result of this up-front group consensus is earlier systems implementation.

Exactly how much time you can save with JAD varies. Companies using JAD typically experience a 40 percent reduction (and sometimes more) in project design time. Here are some comments from companies that have quantified the time savings:

> An application of this complexity could not have been designed and developed [without] the JAD concept. It is estimated that six project months and a cost avoidance of nearly half a million dollars have resulted through the use of JAD.
>
> —Texas Instruments (Cosby, 1985)

> The project leaders for the two Corporate Disbursement System JAD sessions . . . felt they had a four- to six-week time savings.
>
> —(Western-Southern Life, 1986)

> JAD cut out the months of meetings it often took to get users and DP to develop a requirements definition for a proposed system.
>
> —New York Life (Godfrey, 1986)

> A JAD session can consolidate the activities of problem definition, requirements definition, and application design into one elapsed month instead of the usual three to six months.
>
> —(IBM, 1984)

Increase the Quality of the Final Product

JAD brings all the right people together to design systems. An impartial leader guides the participants through a formal agenda. These ingredients produce a product design that is more complete and accurate than could be attained using traditional design methods. Here are some comments from JAD users:

> Participants uncovered things during the JAD that traditional procedures would not have found.
>
> —Prudential (Godfrey, 1986)

> The participants and analysts feel secure in the knowledge that their efforts have been expended in the design of a system to suit the business needs through the 21st century.
>
> —Texas Instruments (Cosby, 1985)

In our own organization, one project manager said the following:

> With JAD, all those who have a major impact on a project have agreed in the session on one direction. No more moving targets! This single-definition approach boosts morale among members of the programming staff because their work is not subject to instant obsolescence after release of the latest "final" specs. We've experienced a high level of programmer commitment to JAD-designed projects. The quality of the finished product is uniformly high.
>
> —John Harkins
> Project Manager, Individual Policy Administration
> Provident Mutual Life Insurance Company

Improve User Relations

A good system design does not guarantee user satisfaction; you can win the battle but lose the war. You can be part of designing the best system of your career but still find that the users are alienated and the programmers won't talk to you.

JAD helps close the traditional communications gap between MIS and users. It creates an atmosphere of problem-solving and cooperation among participants. And it generates group identity. At the beginning of the session, people are reserved, cautious, and even skeptical, especially if JAD has not yet been used in the company. By the end of the session, a spirit of creative enthusiasm has developed, and people have relaxed to the point where, even in the midst of intense technical endeavors, jokes and laughter often punctuate the proceedings.

Again, here is a viewpoint from someone in our organization:

> The JAD methodology has been a breath of fresh air in the develop-
> ment of system requirements. The procedures established to conduct
> the JAD project create an atmosphere of shared responsibility, commit-
> ment, and pride by all participants. This feeling extends well beyond
> the actual JAD meetings. The repeated successes of the JAD methodol-
> ogy have made me a believer. I encourage anyone involved in large-
> scale development to consider introducing the process into your or-
> ganization.
>
> —Karen Senske
> Director, Applications Development
> Provident Mutual Life Insurance Company

The JAD methodology helps MIS participants understand how the
users' business environment works and what they need. In the same way,
users gain a better picture of the agreed-upon design and an appreciation
for work done by MIS. During the sessions, we often hear comments from
user participants like, "so that's how you define data elements" or "no
wonder it takes so long to make changes to the data base." By the end of
the session, participants are generally convinced that JAD is the best way
to design systems, and they often request that it be used for their next
systems project. They are satisfied with the system because *they designed it*.
It is their system. Being committed to its success, they are more willing to
participate in system testing and implementation. As one user said:

> Having attended about five JAD workshops, I am a confirmed advo-
> cate of the methodology. JAD is unparalleled in meeting deadlines
> and getting a new system online.
>
> —George Meholick
> Issue Systems Officer
> Provident Mutual Life Insurance Company

In summary, we have experienced all three of these JAD benefits. The
following is a quote from our Assistant VP, Applications Development:

> Since using JAD at Provident Mutual, we have found that systems go
> into production sooner, user satisfaction has increased, and systems
> require far less modification after installation.
>
> —Roy Clayton
> Assistant VP, Applications Development
> Provident Mutual Life Insurance Company

THE PILOT PROJECT

Of course, it is important that the first JAD project be successful. To help ensure that this pilot project accomplishes your goal, select one that meets the following criteria:

- requires a 12- to 18-month development effort
- involves an online transaction-based system
- is not controversial
- has eager, progressive users who have a strong need for the system
- has strong management, user, and MIS commitment

Nothing can absolutely guarantee the outcome you want on that first project, but selecting a project with these attributes will give you a good chance of success.

TRAINING THE JAD LEADER

JAD success requires a trained JAD leader. Once you have run a few JAD projects, you will need a way to get additional JAD leaders up to speed. There are several companies offering services relating to JAD and other interactive design methodologies. Some of these companies and the services they provide are listed in Figure 13-1.

We feel, however, that on-the-job training is the best way to learn. Training the leader can be done in three steps:

- In the first JAD, the trainee works as an apprentice with an experienced leader. He or she observes all phases of the project. This person attends the interviews and assists in preparing the documents and visual aids.
- In the second JAD, he or she co-leads with an experienced leader, thereby building skills and confidence.
- In the third JAD, the new leader is on his or her own. This person handles all the preparation work and leads the session.

To maintain and further improve your JAD skills, you can establish contacts with other people using the methodology in your area. Some cities have organized JAD user groups. JADLDRS is one such group in the Hartford, Connecticut, area. IBM discusses JAD regularly in their GUIDE user group meetings. No matter how long you have been running JADs, there is always something to be learned from others who also use the methodology.

Company	Method	Services Provided
Boeing Computer Services P.O. Box 33126 Philadelphia, PA 19142	Consensus	Contract session leaders.
IBM Information Services 472 Wheelers Farms Road Milford, CT 06460	JAD	Contract session leaders. JAD training.
JAtech Designer Systems Ltd. 461 Lakeshore Road, West Oakville, Ontario L6K-1G4	JAD	Contract session leaders.
MG Rush Systems, Inc. 21435 Prestwick Barrington, IL 60010	FAST	Contract session leaders. Training in FAST techniques.
Performance Resources, Inc. Five Skyline Place 5111 Leesburg Pike Falls Church, VA 22041	The Method	Contract session leaders. Training and certification in The Method.
Wisdm Corporation 11911 NE 1st Street Suite 206 Bellevue, WA 98005	Wisdm	Contract session leaders.

Figure 13-1 Companies offering interactive design session services

HOW DO YOU MEASURE JAD SUCCESS?

Measuring the success of tools and techniques often comes down to measuring productivity. Some companies count lines of code (LOC). But what does LOC mean?

> Does it mean all the lines a programmer coded during the development of a program (i.e., including throw away code, if any, and the code that was added, changed, or deleted because of specification changes), or just the lines in the final product? Does it include data definitions and comments, or just the executable statements? Unless LOC is precisely defined, the productivity measurements do not make much sense and the productivity comparisons are misleading. (Parikh, 1982)

Others use function point analysis, where the total amount of function provided by the system is quantified by counting external inputs, outputs, interfaces, inquiries, and logical files of an application. And still others compare work volumes to resources using such formulas as the following. (With this, the larger the result, the more productive the project.)

$$\text{Productivity} = \frac{\text{Work product}}{\text{Work effort}}$$

While researching a report on COBOL preprocessors, Girish Parikh asked several professionals for information on increasing productivity. One reply was, "The improvement in programmer productivity has not been measured statistically. Most shops do not measure productivity at all—they just have a gut feeling about it. . . ." (Parikh, 1982)

In the same way, we feel that JAD success cannot be determined through formulas or counting lines of code. The real evaluation comes down to this:

- *Work product*. Based on the final JAD Design Document, can the MIS team take the next step in the systems development life cycle?

- *Commitment*. Do both the users and MIS have a complete picture of the design and do they agree with it? Do they want it to succeed? Will the users continue their involvement during implementation?

- *Enthusiasm*. Do the users feel the design will help them perform their jobs more effectively than before? Do the participants (users and MIS) want to use JAD again?

You may have other criteria for measuring success. For example, in a survey given by the Royal Bank of Canada, JAD success was measured strictly on user satisfaction. (Brown, 1987) Figure 13-2 shows the results.

JAD User Satisfaction Survey

1. Do you feel that the requirements/specifications that resulted from the session more accurately reflected those of the user than if they had been prepared in the more traditional manner?

Not at all				Much more accurate
1	2	3	4	5
			74%	26%

2. Do you feel any improvement in communications resulted from the session?

Not at all				Much improved
1	2	3	4	5
	11%	16%	42%	31%

3. Do you feel that the requirements/specifications were completed in less elasped time than they would have been using traditional methods?

Not at all				Much sooner
1	2	3	4	5
			53%	47%

Figure 13-2 JAD user satisfaction survey

CONCLUSION

We have described for you the JAD methodology—its phases, psychology, tools, and techniques. You have read the case study and, at the same time, know that no two JADs are alike. You understand the importance of impartiality on the part of the JAD leader, of having the right people in the session, and of following a structured agenda.

Now the question is this: Can you do it? Can you stand in front of 15 people for three days and guide them through the agenda for designing a new system? Once you have management support for using the methodology, the only person left to persuade is yourself. The only thing standing between you and a stream of successful JADs is your own uncertainty. You can get through that only by jumping in and doing it. Then, after a couple of successful projects, your management, your users, and the MIS staff will understand what the methodology can accomplish. They will see JAD as an efficient way to design quality systems in a shorter time. They will see the power of JAD.

APPENDIX A

Appendix A shows the nine scribe forms used to capture information for the Management Definition Guide. In Phase 1 of the JAD project, the JAD leader gives these forms to the key user managers who research the requirements and complete the forms. The resulting information is compiled into the Management Definition Guide.

PURPOSE OF THE SYSTEM

_____

_____

_____

_____

_____

_____

_____

_____

_____

_____

_____

_____

_____

_____

_____

_____

_____

_____

_____

SCOPE OF THE SYSTEM

Who uses the system and how often:

_____

_____

_____

_____

_____

Expected growth of business:

_____

_____

_____

_____

_____

Areas providing source documents or receiving reports:

_____

_____

_____

_____

_____

System interfaces:

_____

_____

_____

_____

_____

MANAGEMENT OBJECTIVES

1. _____
   _____

2. _____
   _____

3. _____
   _____

4. _____
   _____

5. _____
   _____

6. _____
   _____

7. _____
   _____

8. _____
   _____

9. _____
   _____

10. _____
    _____

FUNCTIONS

1. _____
   _____
2. _____
   _____
3. _____
   _____
4. _____
   _____
5. _____
   _____
6. _____
   _____
7. _____
   _____
8. _____
   _____
9. _____
   _____
10. _____
    _____

CONSTRAINTS

1. _____
   _____
   _____

2. _____
   _____
   _____

3. _____
   _____
   _____

4. _____
   _____
   _____

5. _____
   _____
   _____

6. _____
   _____
   _____

ADDITIONAL USER
RESOURCE REQUIREMENTS

PEOPLE _____
_____
_____
_____
_____

PHYSICAL SPACE _____
_____
_____
_____

HARDWARE _____
_____
_____
_____
_____

OTHER EQUIPMENT _____
_____
_____
_____

ASSUMPTION
(Pre-Session Version)

SUBJECT _____

DESCRIPTION _____

_____

_____

_____

_____

_____

_____

_____

_____

_____

_____

_____

_____

_____

_____

_____

OPEN ISSUE
(Pre-Session Version)

ISSUE NAME _____

DESCRIPTION _____

_____

_____

_____

_____

_____

_____

_____

_____

_____

_____

_____

_____

_____

PARTICIPATION LIST

Who	Department	Mail Code	Role

APPENDIX B

Appendix B shows the scribe forms used to document decisions made in the JAD session. The information on these forms is compiled into the final JAD Design Document.

You can copy these forms for your own JAD projects or modify them to meet your needs.

ASSUMPTION

SUBJECT _____

SOURCE (Enter the agenda topic - for example, screens - or open issue number from which the assumption originated)
_____

DESCRIPTION _____

_____

_____

_____

_____

_____

_____

_____

_____

_____

_____

_____

_____

_____

WORK FLOW DIAGRAM

WORK FLOW TITLE _____

DATA ELEMENT DESCRIPTION

DATA ELEMENT _____

LENGTH _____

FORMAT _____

DESCRIPTION _____

_____

_____

_____

_____

_____

_____

_____

_____

_____

_____

_____

SCREEN FLOW DIAGRAM

SCREEN FLOW _____

SCREEN DESCRIPTION

SCREEN NAME _____

DESCRIPTION _____

_____

_____

_____

_____

_____

_____

_____

_____

_____

_____

_____

_____

_____

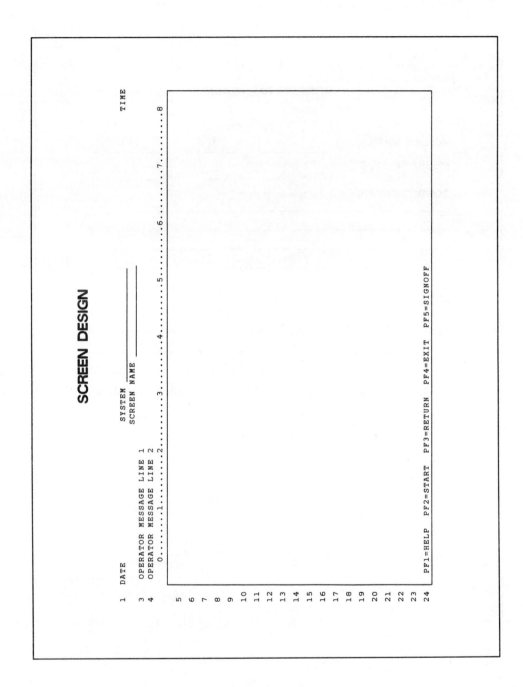

SCREEN MESSAGE

SCREEN NAME _____

MESSAGE TYPE
Confirmation _____   Error _____   Other _____

CONDITIONS (When does this message display?)

_____

_____

_____

_____

_____

MESSAGE TEXT _____

_____

_____

_____

_____

_____

_____

FIELDS BY SCREEN

Screen Name _____

B=Bright I=Inquiry N=Normal U=Update					
Screen Field Name	B/N	I/U	*Source*	*Target*	*Screen Field Values*

SCREEN ACCESS BY JOB FUNCTION

Screen Name ----------- >										
Job Function	*No. of Terminals*									

I = Inquiry
U = Update
X = Either

REPORT DESCRIPTION

REPORT NAME _____

DESCRIPTION _____

_____

_____

_____

_____

FREQUENCY _____

COPIES _____

DISTRIBUTION _____

SELECTION _____

SORT BY _____

DATA ELEMENTS

_____   _____

_____   _____

_____   _____

_____   _____

_____   _____

REPORT DESIGN

REPORT TITLE SYSTEM NAME _____

PAGE XX

```
 1  DATE
 2
 3
 4
 5
 6
 7
 8
 9
10
11
12
13
14
15
16
17
18
19
20
21
22
23
24
25
26
27
28
29
30
```

REPORT LIST

1. _____
2. _____
3. _____
4. _____
5. _____
6. _____
7. _____
8. _____
9. _____
10. _____
11. _____
12. _____

OPEN ISSUE

ISSUE NO. _____

ISSUE NAME _____

ASSIGNED TO _____

_____

_____

RESOLVE BY (Date) _____

DESCRIPTION _____

_____

_____

_____

_____

_____

_____

_____

_____

_____

_____

_____

RECORD DESCRIPTION

RECORD NAME _____

DESCRIPTION _____

_____

_____

_____

_____

_____

DATA ELEMENT NAMES

_____ _____

_____ _____

_____ _____

_____ _____

_____ _____

_____ _____

_____ _____

_____ _____

_____ _____

_____ _____

_____ _____

RECORD DESIGN

(based on Bachman diagram)

Record Name			
Record ID	Record Format	Record Length	Location Mode
CALC Key Name or VIA Set Name			Dups Option
Area Name			

RECORD VOLUME

Record volume is the number of occurrences anticipated for each record type.

Record Name	Expected Volume	Maximum Volume

SET LENGTH

Set length is the number of member occurrences per owner occurrence.

Set Name	Expected Length	Maximum Length

LOGICAL DESIGN

DATA BASE NAME _____

TRANSACTION

TRANSACTION _____

SYSTEM _____

DESCRIPTION _____

_____

_____

_____

_____

_____

DATA ELEMENT NAMES

_____   _____

_____   _____

_____   _____

_____   _____

_____   _____

_____   _____

_____   _____

_____   _____

CALCULATION ROUTINE

CALCULATION _____

DESCRIPTION _____

_____

_____

_____

_____

_____

_____

_____

_____

_____

_____

_____

_____

_____

MANUAL FORM DESCRIPTION

FORM NAME _____

DESCRIPTION _____

_____

_____

_____

_____

_____

_____

_____

_____

_____

_____

_____

MANUAL FORM DESIGN

FORM NAME _____

APPENDIX C

Appendix C shows the sample Management Definition Guide prepared in the Fixed Assets case study described in Chapter 11. Compiled before the JAD session, this document contains management's view of the purpose, scope, and objectives of the system.

MANAGEMENT DEFINITION GUIDE

FIXED ASSETS

October 19, 1987

 BUSINESS SYSTEMS ENGINEERING

PREFACE

> *This document results from the first phase (Project Definition) of the JAD methodology. Information gathered in interviews with the executive sponsor, key users and ISD managers has been compiled into this document, the Management Definition Guide.*

TABLE OF CONTENTS

PURPOSE OF THE SYSTEM

The purpose of the Fixed Assets system is to provide accounting and reporting information for the Finance department to track the capital assets purchased for the company and its subsidiaries. For this project, fixed assets includes those items that:

- have useful lives greater than 12 months
- satisfy capital requirements as determined by asset classification
- are intended to be held in the operating activity of the business
- are tangible, that is, have physical substance

SCOPE OF THE SYSTEM

The Fixed Assets system will include:

- Home Office Furniture
- Home Office Equipment
- Field Furniture
- Field Equipment
- EDP Equipment
- Leasehold Improvements
- Multi-Company Processing

The following areas of the company are affected either through direct system interaction or indirectly by providing source documents or receiving reports:

- Budget and Cost
- Facilities Administration
- Information Services Department (ISD) Planning
- Marketing Properties
- Purchasing
- Tax Planning

The Fixed Assets system will interface with:

- CDS (Cash Disbursement System) for acquisitions
- ABC (Accounting Budget and Cost) for depreciation, chargeback, and disposal accounting of Home Office Furniture; and unit cost accounting and chargeback of Field Furniture
- AER (Automated Expense Redistribution) for EDP equipment chargeback

MANAGEMENT OBJECTIVES

Management objectives for the Fixed Assets system are as follows:

1. Perform online data entry, inquiry, edits, and updates of fixed assets data

2. Record the acquisition, depreciation, and disposition of fixed assets to conform with STAT/GAAP and Tax guidelines

3. Maintain detail data on asset purchases, dispositions, and transfers to support the ABC ledger

4. Compute proper book and tax depreciation

5. Reconcile with the company's trial balance asset accounts

6. Compute monthly cost center chargebacks

7. Provide the basis for determining gain or loss on the disposition of fixed assets

8. Maintain various depreciation calculations and balances for each asset

9. Compute depreciation using various methods

10. Provide the following reports:

- standard reports that include transaction registers, depreciation summaries, and tax information
- control reports that summarize asset activity, provide audit trail controls, and balance files
- ad hoc reports via a flexible report writer

11. Microfiche all reports

12. Provide year-to-date historical transaction file

13. Provide complete system documentation

14. Provide security and access privileges at the screen level

FUNCTIONS

This section has been abbreviated to include functions only for Home Office Furniture, Field Furniture, and Multi-Company Processing.

Home Office Furniture

The system will provide the following functions for Home Office Furniture:

1. Add, dispose, and transfer inventory items (including partial and complete)

2. Display information:
   a. Items by category ID
   b. Items by cost center
   c. Category table
   d. Item record

3. Update depreciation information:

   a. Handle both tax and statutory information
   b. Calculate monthly depreciation
   c. Update item records
   d. Feed accounting entries to ABC

4. Update monthly chargeback information:

   a. Calculate monthly rate on a yearly basis
   b. Calculate monthly chargeback for cost centers
   c. Update item records
   d. Feed accounting entries to ABC

5. Calculate gain/loss for statutory and tax reporting
6. Track expensed items
7. Do mass transfers and disposals at the category and cost center levels
8. Update category, depreciation, and account number tables
9. Do error correction and reversal accounting
10. Create management, control/audit trail, and tax reports

Field Furniture

The system will provide the following functions for Field Furniture:

1. Add, dispose, transfer, change, and correct inventory items (including partial and complete)
2. Display information:

   a. Items by category ID
   b. Items by cost center
   c. Category table
   d. Item record

3. For unit cost reporting:

   a. Do amortization calculations
   b. Feed accounting entries to ABC

4. Identify items by location and distinguish among different locations of the same agency
5. Do mass transfers and disposals at the category, location, and cost center levels
6. Update category and account number tables
7. Do error correction and reversal accounting
8. Create management, control/audit trail, and tax reports

9. Calculate tax depreciation
10. Calculate gain and loss for tax reporting

Multi-Company Processing

The system will provide the following functions for Multi-Company Processing:

1. Provide detail records by company
2. Update tables separately (for example, depreciation and chargeback tables)
3. Sort reports and subtotal by company

CONSTRAINTS

Constraints have not been defined for this preliminary Request for Proposal (RFP) phase.

ADDITIONAL USER RESOURCE REQUIREMENTS

Since the Finance department will take on the work that is currently handled by several departments (see Assumption number 2), they will need additional staff. The actual number will be determined after user requirements have been defined.

ASSUMPTIONS

The design group should make the following assumptions about the business and operating environment:

1. Purpose of the JAD.
    The purpose of this JAD is to determine system requirements that will be used:

    *   as part of the RFP to submit to outside vendors
    *   by ISD to estimate the cost of developing the system in-house

    If the users decide to have the system developed in-house, another JAD would be scheduled to define the detailed system requirements. Since the purpose of this JAD does not involve the actual detailed design, we will *describe* the screens and reports but not *design* their formats.

2. Centralization in Finance.
    The work will be centralized in the Finance department:

294 JOINT APPLICATION DESIGN

- For all fixed assets in the company, Finance will reconcile the detail records in the Purchasing ledger as well as handle payments. Purchase decisions will continue to be made in their related areas.
- All data entry (other than Information Services Operations) will be done in Finance for assets purchased by the company. Other departments using the current system will be able to inquire into the system. Tax Planning will be able to approve tax depreciation and useful life.
- The Finance departments of subsidiaries that purchase their own fixed assets will update their own fixed assets information.
- Leasehold improvements are currently handled on a Lotus 1-2-3 program for depreciation in the Mortgage Loan and Real Estate department. This will also move to Finance.
- Finance will be staffed to handle the new Fixed Assets system work flow. The needs will be analyzed after user requirements have been defined in the JAD session.

3. Archival Requirements.
To maintain a history of fixed assets information:

- Save all transactions for a designated period of time.

4. Out-of-Service Items.
Inventory items that are out of service (but ready to use) will continue to be charged to the cost centers as they are today. For example, out-of-service Memorex terminals will continue to be charged to ISD. A unique identifier will be attached to items to identify them as stock.

OPEN ISSUES

> *This section has been abbreviated by eliminating some open issues.*

1. Vehicles and Buildings.
Will the system handle fixed assets information for the company's fleet of automobiles?

2. EDP Chargeback Rates.
Will the Fixed Assets system calculate EDP chargeback rates? If not, will the system feed the rates to the EDP chargeback system?

3. Use of the Fixed Assets System to Track Inventory.
Should the Fixed Assets system also track inventory for all or certain classes of assets?

4. Home Office Furniture Inventory Methodologies.
Should furniture for the home office be tracked by workstations or by component part?

5. Fixed Assets and Unit Cost Interface.
   Is there an interface between the Fixed Assets system and the unit cost program for Field Furniture? If so, should this be automated?

6. Capitalization Thresholds.
   Should the current capitalization thresholds be raised to accommodate new tax laws?

7. Mass Purchases.
   Should mass purchases be capitalized? This refers to items of a small unit purchased in bulk (in excess of $5,000, for example).

8. Partial Updates.
   What is involved in a partial update including calculating partial depreciation and rental? What is the impact on other areas? How will the system handle adding, transferring, and disposing items?

9. Projected Depreciation.
   Should the system be able to project depreciation in future periods for budget or financial statement purposes?

Name	Job Title	Mail Code	Role
David Bland	Marketing Properties	8	Team
Jessica Brooke	Tax Planning and Research	9	Team
Hugh Close	Tax Planning and Research	9	Team
Carol Davis	ISD, Automation Planning	34	Team
Virginia Gamby	Accounting & Budget	7	Team
Jane Grais	ISD, Applications Development	34	Team
Barbara Haber	Budget & Cost	7	Team
Mark Hellman	Accounting Systems	7	Team
Jesse Honey	Accounting Systems	7	Team
Stella Kowalski	Financial Reporting & Planning for Non-Insurance Subsidiaries	7	Team
Bruce Lee	Administrative Services	5	Team
Michael Marks	ISD, Applications Development	34	Team
Phil Mugler	ISD, Applications Development	34	Scribe
Bill Platt	Agency	8	Team
Josephine Rose	Controller	7	Executive Sponsor
Robin Rosen	Space Planning and Facilities Mgt.	8	Team
Denise Silver	ISD, Business Systems Engineering	34	Leader
Jane Wood	ISD, Business Systems Engineering	34	Leader

Figure 1 JAD session participants

10. Fully Depreciated Assets.
    Should fully depreciated assets come off the Fixed Assets system and go into a history file?

11. Audit Trails.
    Should an audit trail be produced that includes the before and after changes in the data base for any asset?

JAD SESSION PARTICIPANTS

The participants for the JAD session will be as listed in Figure 1, on page 295.

APPENDIX D

Appendix D shows the sample JAD Design Document prepared in the Fixed Assets case study described in Chapter 11. This document contains the specifications agreed upon in the JAD session.

JAD DESIGN

FIXED ASSETS

November 6, 1987

David Bland
Jessica Brooke
Hugh Close
Carol Davis
Virginia Gamby
Jane Grais
Barbara Haber
Mark Hellman
Jesse Honey
Stella Kowalski
Bruce Lee
Michael Marks
Phil Mugler
Bill Platt
Josephine Rose
Robin Rosen
Denise Silver
Jane Wood

 BUSINESS SYSTEMS ENGINEERING

PREFACE

This design document describes user requirements for the Fixed Assets system. It includes all specifications defined in the JAD session held October 26 to 29, 1987. When the participants approve this document, the Finance department can begin preparing a Request for Proposal (RFP) to send to the vendors and ISD can begin estimating the cost of developing the system in-house.

TABLE OF CONTENTS

LIST OF ILLUSTRATIONS

JAD OVERVIEW

Definition

Joint Application Design (JAD) is a methodology for defining application requirements. It brings together users and the Information Services Department (ISD) in a workshop conducted by a leader. The methodology culminates in a comprehensive document containing the specifications agreed upon by all the participants. JADs are conducted by the Business Systems Engineering (BSE) unit.

Benefits

The JAD methodology does the following:

- accelerate system design
- increase the quality of the final product
- improve user and ISD relations

JAD Criteria

The JAD methodology is used for projects that:

- affect more than one department
- require at least six ISD person-months to implement
- have high business priority or complex requirements

Open Issues and Assumptions

Throughout the project, questions can arise regarding user requirements. When these questions cannot be resolved, they are designated as *open issues*. Some are resolved before the session and are added to a list of *assumptions*. Others remain unresolved and are carried into the session for discussion.

During the session, some issues are resolved while new ones are identified. At the close of the session, all unresolved issues are assigned to the appropriate personnel. A resolve data is determined and a coordinator is designated for issues assigned to more than one person.

After the session, when an issue is resolved, the issue coordinator sends a written copy of the resolution to the executive sponsor and to all participants.

JAD Design Document

The *JAD Design Document* is the final product of the JAD project. It represents agreements reached by the session participants. After the user and ISD managers approve the document, it is turned over to Applications Development, where they begin the program design phase.

As specifications evolve after the document is produced, BSE does *not* change the contents of the final document. However, the document files can be copied for the programmers to maintain. The original files remain with BSE.

The JAD Phases

The JAD methodology consists of these five phases:

Phase	Description	Resulting Output
1	Project Definition	Management Definition Guide
2	Research	Work flow Preliminary specifications JAD session agenda
3	Preparation	Working Document Overheads, flip charts, magnetics
4	The JAD Session	Completed scribe forms
5	The Final Document	JAD Design Document Signed approval form

The following describes each phase.

1. Project Definition.
   The JAD leader, from BSE, interviews the executive sponsor, key users, and ISD project managers. The purpose, scope, and objectives of the project are defined. This information is compiled into a document, called the *Management Definition Guide*, which is distributed to all participants before the session. The scribe and session participants are selected and the session dates are determined.

2. Research.
   The leader interviews user and ISD managers and others familiar with the system. The work flow is documented and preliminary specifications are gathered for data elements, screens, and reports. The JAD session agenda is determined.

3. Preparation.
   The work flow and other preliminary specifications are combined to produce the Working Document. This is distributed to all participants and used as the basis for the session. At the same time, BSE prepares visual aids and trains the scribe.

4. The JAD Session.
   During the session, the leader directs the participants in defining and documenting the user requirements. The scribe documents decisions on

scribe forms. Open issues are resolved or assigned to be resolved after the session.

5. The Final Document.
   The completed scribe forms are used to produce the final JAD Design Document. This complete synthesis of user requirements is distributed to all participants. A review meeting is held. When all participants agree on the contents, the document is approved.

JAD Roles

The following describes the roles and responsibilities of the JAD participants.

Business Systems Engineering (BSE)

BSE performs activities associated with developing and enhancing major automated systems. This can include the following tasks within the systems development life cycle:

- user requirements definition (accomplished through JAD)
- acceptance testing
- user training and documentation

Executive Sponsor

Management commitment is important to the success of the system design. The executive sponsor is a user with senior-level authority to make decisions for all aspects of the project. This person is not necessarily an active participant during the session itself. Responsibilities include the following:

- Before the session, the executive sponsor provides a management perspective and defines the purpose, scope, and objectives of the effort. This is documented in the Management Definition Guide.
- During the session, this person handles any open issues that bring the session to an impasse.
- After the session, this person monitors the resolution of open issues.

JAD Leader

JAD leaders from BSE serve as impartial session moderators. Each session has one leader and a BSE support person who remain with the project throughout all the JAD phases. Their responsibilities include the following:

- Before the session, they conduct interviews with key users and ISD personnel to become familiar with the current system and gather specifications for the new system. They document work flow using data flow diagrams.

They compile all this information into the Working Document. And they prepare visual aids such as overheads, flip charts, and magnetics to graph-ically support the discussion.

- During the session, the leaders guide the participants through the agenda, assuring that both user and ISD interests are represented. They work with the scribe to document the specifications agreed upon during the session.
- After the session, the leaders coordinate the production, review, and dis-tribution of the JAD Design Document.

Scribe

The scribe documents the decisions made in the session. This person is selected from the user area or from ISD. Responsibilities include the following:

- Before the session, the scribe works with the leader to become familiar with the JAD methodology and the scribe forms used in the session.
- During the session, the scribe uses the Working Document and the scribe forms to document the agreed-upon specifications. These completed forms along with the edited Working Document are the basis for the JAD Design Document. Sometimes, the scribe generates new and modified screens or reports for the session the next day. Although the sessions are half-day, the scribe may be required to work a full day to assist in JAD-related tasks.
- After the session, based on project needs, the scribe might help generate specific pages for the JAD Design Document.

Full-Time Participants

Full participation of the people in the session is critical to the success of the project. This means that those who have been designated as full-time participants should attend the entire session. The participants include users and ISD per-sonnel.

Users range from lead clerks to upper management. This diversity helps ensure that all subject areas can be covered in detail. Since the users are develop-ing the application solution, it is essential that as a group they have complete knowledge of the system and the authority to make decisions on the operational and functional aspects of the system.

ISD participants range from programmers to project leaders. They provide background information on existing systems and technology. Also, ISD participa-tion ensures that they will understand user needs and be able to translate the resulting JAD Design Document into an effective system design.

The responsibilities of user and ISD participants include the following:

- Before the session, they participate in documenting the existing work flow and prototyping the new work flow. They provide samples of existing and prototype screens and reports, and gather other information relating to the system design to add to the Working Document.

- During the session, they participate in discussions to define user requirements.
- After the session, they review the JAD Design Document.

Users on Call

These people do not attend the entire session but are present for certain topics
in the agenda where their expertise is required.

AGENDA

This section includes a discussion of the following:

- session agenda
- session participants
- distribution for the final document

Session Agenda

The following agenda was accomplished in the four-day session held October 26
to 29, 1987.

1. Existing work flow
2. Assumptions
3. Current open issues
4. New work flow
5. Screens
6. Processing (calculations and reversal processing)
7. Reports
8. New open issues
9. Distribution list for the final document

Session Participants

The participants for the JAD session are shown in Figure 1.

Distribution for the Final Document

The final JAD Design Document will be sent to the people listed in Figure 2.

Name	Job Title	Mail Code	Role
David Bland	Marketing Properties	8	Team
Jessica Brooke	Tax Planning and Research	9	Team
Hugh Close	Tax Planning and Research	9	Team
Carol Davis	ISD, Automation Planning	34	Team
Virginia Gamby	Accounting & Budget	7	Team
Jane Grais	ISD, Applications Development	34	Team
Barbara Haber	Budget & Cost	7	Team
Mark Hellman	Accounting Systems	7	Team
Jesse Honey	Accounting Systems	7	Team
Stella Kowalski	Financial Reporting & Planning for Non-Insurance Subsidiaries	7	Team
Bruce Lee	Administrative Services	5	Team
Michael Marks	ISD, Applications Development	34	Team
Phil Mugler	ISD, Applications Development	34	Scribe
Bill Platt	Agency	8	Team
Josephine Rose	Controller	7	Executive Sponsor
Robin Rosen	Space Planning and Facilities Mgt.	8	Team
Denise Silver	ISD, Business Systems Engineering	34	Leader
Jane Wood	ISD, Business Systems Engineering	34	Leader

Figure 1 Session participants

ASSUMPTIONS

This section includes all basic business decisions that have been made about the Fixed Assets system project. These assumptions should be kept in mind while designing the system.

1. Purpose of the JAD.
   The purpose of this JAD was to determine system requirements that will be used (1) as part of the Request for Proposal (RFP) to submit to outside vendors; and (2) by ISD to estimate the cost of developing the system in-house. If the users decide to have the system developed in-house, another JAD would be scheduled to define the detailed system requirements. Since the purpose of this JAD did not involve the actual detail design, we will describe the screens and reports but not design their formats.

2. Centralization in Finance.
   The work will be centralized in the Finance department as follows:

   • For all fixed assets in the company, the Finance department will reconcile the detail records in the Purchasing ledger as well as handle payments. Purchase decisions will continue to be made in their related areas.

Name	Copies	Mail Code
David Bland	1	8
Jessica Brooke	1	9
Hugh Close	1	9
Barbara Dash	1	9
Carol Davis	1	34
Sally Doehnert	3	34
Ralph Garlitos	1	8
Jane Grais	1	34
Barbara Haber	3	7
Mark Hellman	1	7
Joan Holl	2	5
Jesse Honey	1	7
Jenna Kauffman	1	8
Stella Kowalski	1	7
Bruce Lee	1	8
Michael Marks	1	34
Ned Mazer	1	9
Phil Mugler	1	34
Bill Platt	1	8
Bob Potter	1	34
Josephine Rose	1	7
Robin Rosen	1	8
Denise Silver	1	34
Jane Wood	1	34

Figure 2 Distribution for the final document

- All data entry (other than Information Services Operations) will be done in Finance for assets purchased by the company. Other departments using the current system will be able to inquire into the system. Tax Planning will be able to approve tax depreciation and useful life and correct figures if they are wrong.
- The Finance departments of subsidiaries that purchase their own fixed assets will update their own fixed assets information and upload it to the parent company.
- Leasehold improvements (alterations or improvements of leased premises) are currently handled on a Lotus 1-2-3 program for depreciation in the Mortgage Loan and Real Estate department. This function will move to Finance. Leasehold improvements should be amortized over the life of the lease or the useful life of the improvements, whichever is shorter.
- Finance will be staffed to handle the new Fixed Assets work flow. The needs will be analyzed after user requirements have been defined in the JAD session.

3. Home Office Furniture Inventory Methodologies.
The participants evaluated the following inventory methods to use for Home Office Furniture:

- Method A (Inventory by Workstation). This is the method currently used where components are tracked by workstations, for example, "A" stations, "B" stations, and so on.

- Method B (Inventory by Component Part). For this method:
  - inventory will be tracked by component part at the company level
  - chargeback will be done by the ABC system
  - each responsibility center will have a chargeback amount based on a more general formula (for example, square footage or staffing) as opposed to keeping a specific inventory by cost center.

Figure 3 shows the participants' evaluation of Methods A and B.

- The first column, *Criteria*, shows the criteria identified by the team for evaluating the two methods.
- The second column, *Weight Factor*, shows the relative importance given to each criterion. The weights are 1 through 3, where 3 is the most important.
- The last two columns show Methods A and B.
  - The *Score* columns show the relative rating, where 4 is the best. For example, the criterion "Ease of system construction" rated fairly high (3) for Method A and low (1) for Method B.
  - The *Score × Weight* columns show the numbers resulting from multiplying the Score by the Weight Factor. This gives a relative weighted rating where *the higher the number, the better the rating*.

Based on this evaluation, Method B is the best approach.

4. Items and Categories.
   The system should handle depreciation calculations at the item level. The depreciation information (such as depreciation period, depreciation method, and date fully depreciated) should be calculated for each item rather than in groups within the category table (as it was before). Users will be able to change the information in the item record. Figure 4 defines

Criteria	Weight Factor	Method A Fixed Assets by Workstation		Method B Fixed Assets by Component Part	
		Score	Score × Weight	Score	Score × Weight
Accuracy of the subsidiary ledger	3	1	3	3	9
Work vs. return on investment	2	1	2	3	6
Chargeback accuracy	2	3	6	1	2
Cost control	1	3	3	2	2
Ease of system construction	1	3	3	1	1
Ease of system conversion	1	3	3	4	4
Totals			20		24

Figure 3 Evaluation of inventory methods for Home Office Furniture

	Home Office Furniture	Field Furniture	EDP Equipment	Home Office Equipment	Field Equipment	Leasehold Improvements
Item Examples	For method A: freestanding furniture, prints, workstations For method B: filing cabinets, prints	Freestanding furniture	Memorex terminals with specific serial nos.	calculators, typewriters, telephone equip.	calculators, typewriters	carpeting, electrical service, wiring, ceiling tiles, water fountains, vinyl wall coverings
Identified by:	category ID	category ID	serial no.	serial no., voucher number for telephone equipment	serial no.	category ID, building location
Category examples	For method A: art work, task seating, guest seating For method B: parts & pieces	armchair, bookcase	terminals, printers, CPU disk, PC cables	elec. typewriter, std. typewriter	elec. typewriter, std. typewriter	flooring, lighting

Figure 4 Items and categories for fixed assets groups

items and categories in terms of how they will be identified for the various fixed assets groups. For Home Office Furniture, examples are given for both Methods A and B.

5. Fixed Assets as an Inventory System.
   The system will track inventory for all classes of assets. For Home Office Furniture, if inventory is recorded at the component level (Method B), it will not be tracked by cost center.

6. Tax Requirements.
   The system will provide the following information for tax purposes:

   • Each item of inventory and each acquisition must have an associated purchase date, year placed in service, and purchase price.

   • The system must either agree with the trial balance or be reconcilable.

   • No item will be taken off the inventory system until it is disposed of (through sales, trade-ins, donations, and so on) for both statutory (useful life as set by government regulations) and tax. This requires separate retirement functions for statutory and tax.

   • Report requirements are listed in the Reports section.

7. Separate Capitalization Thresholds.
   The system will be able to handle separate capitalization thresholds for statutory and tax records.

8. Items That Fall Below the Capitalization Minimum.
   Items that fall below the capitalization minimum ($50) will be expensed (as a write-off) and charged back to the user area in the month of purchase.

For these situations, categories (such as PC boards and cables) will be determined.

9. Projected Depreciation.
   For existing assets, the system will be able to project total depreciation in future periods for budget and financial statements.

10. Chargeback Rates.
    EDP chargeback rates will not be calculated by the Fixed Assets system. Instead, the chargeback rate will be manually loaded into the system. Then the actual monthly chargeback amount for each cost center will be automatically fed to the Chargeback system. The chargeback rate will be included in the Category Table.

11. Mass Purchases (Volume Buying).
    Mass purchases will be input to an *inventory* cost center and transferred to a *location* cost center where the items will be placed. Thus the value of the original inventory cost center will be reduced while the new location cost center will be increased.

12. Out-of-Service Items.
    Inventory items that are out of service (but ready to use) will continue to be charged to the cost centers as they are today. For example, out-of-service Memorex terminals will continue to be charged to ISD. A unique identifier will be attached to items to identify them as stock.

13. Disposing Assets with No Unique Identifier.
    If an item can be identified by year of purchase, then that particular item will be disposed. Otherwise, the FIFO (first in first out) method will be used. (The item with the earliest purchase date will be disposed first.)

14. Vehicles and Buildings.
    The system will handle fixed assets information for vehicles and buildings.

15. Audit Trails.
    Audit trails will be produced that include the before and after changes in the data base for all assets.

16. Amortization Rates.
    The system will include a separate amortization table for Field Equipment and Field Furniture. Users will be able to determine whether or not to amortize at the item level.

17. Archival Requirements.
    To maintain a history of fixed assets information:

    - Save all transactions until the tax year is closed.
    - Microfiche all reports.
    - Track all items in inventory. Delete disposed items from the data base and archive them for future reference.

18. Fully Depreciated Assets.
    Fully depreciated assets should *not* come off the Fixed Assets system and go into a history file.

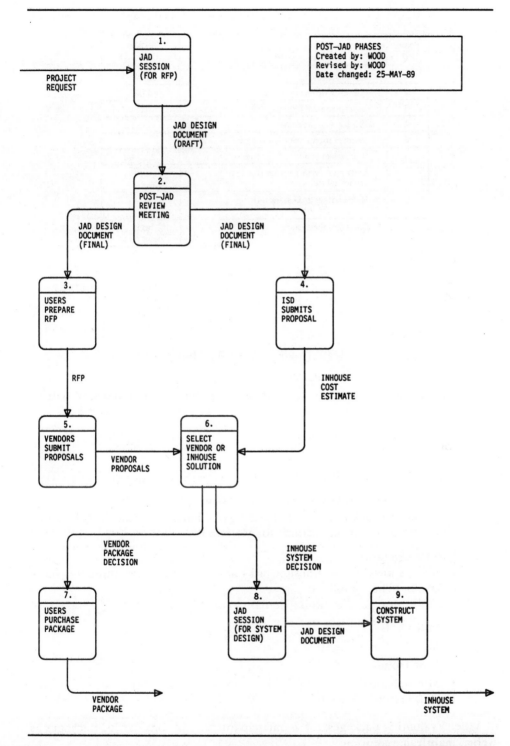

Figure 5 Post-design tasks

Phase No.	Phase	Date Complete
1	JAD session	10/29/87
2	Post-JAD review meeting	11/87
3	Users prepare RFP	12/15/87
4	ISD submits proposal	1/15/88
5	Vendors submit their proposals	–
6	Users select vendor or in-house solution	2/29/88
7	Users purchase a package (for the vendor solution) or	–
8	Hold another JAD session (for the in-house solution)	–
9	Construct the system	to be determined

Figure 6 Dates for the post-design tasks

19. Items to Estimate Separately.
    For the in-house ISD cost estimate, provide separate estimates for:

    * automated accounting for reversals
    * automated calculation for change in depreciation period or end date
    * interfaces
    * conversion requirements
    * accounting for leases

    Separate estimates are not needed for:

    * ad hoc reporting (since ISD will not create those reports)
    * the chargeback portion of Home Office Furniture (since it will not be processed on the mainframe for Method B)

20. Post-Design tasks.
    Figure 5 shows the post-design tasks and Figure 6 shows the dates these tasks should be completed.
    The following people will participate in the RFP process:

    * Jane Grais
    * Mark Hellman
    * Jesse Honey
    * Michael Marks
    * Carol Davis

When evaluating packages, these participants should consider PC packages as well as mainframe solutions.

WORK FLOW

This section includes data flow diagrams for:

- existing work flow
- new work flow
- the interface between the Cash Disbursement and Fixed Assets systems

> *This section has been abbreviated. The new work flow is not included.*

Existing Work Flow

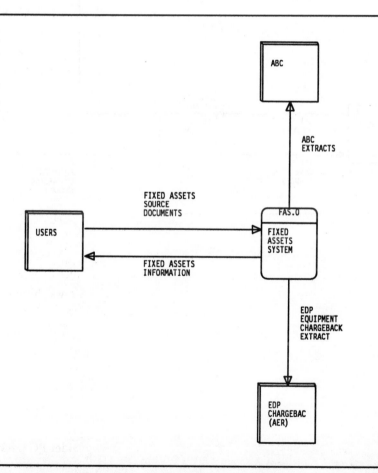

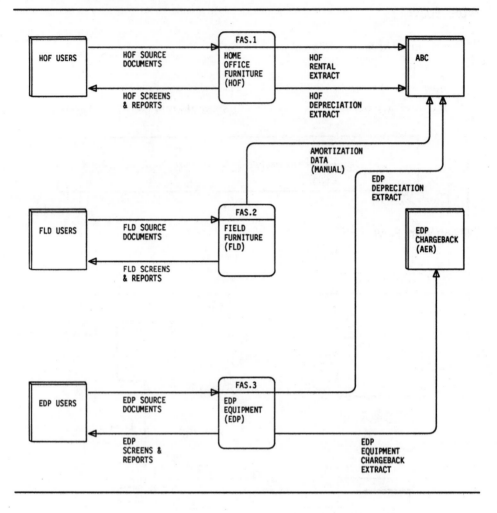

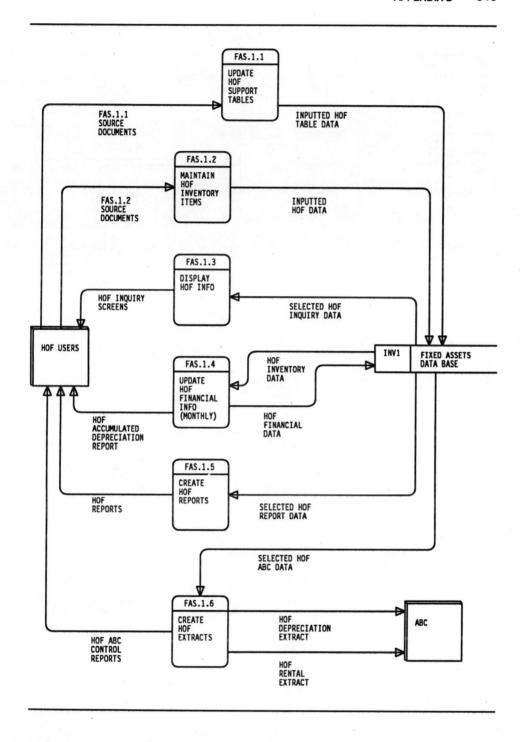

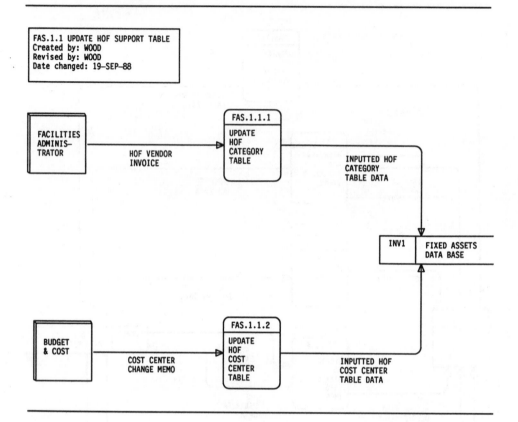

FAS.1.1 UPDATE HOF SUPPORT TABLE
Created by: WOOD
Revised by: WOOD
Date changed: 19-SEP-88

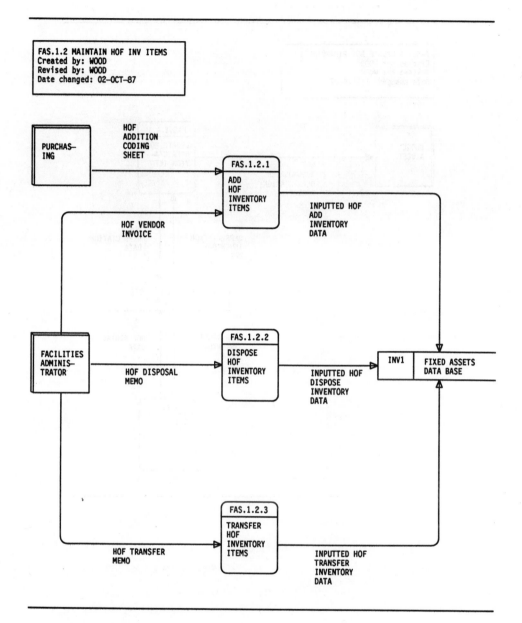

FAS.1.2 MAINTAIN HOF INV ITEMS
Created by: WOOD
Revised by: WOOD
Date changed: 02—OCT—87

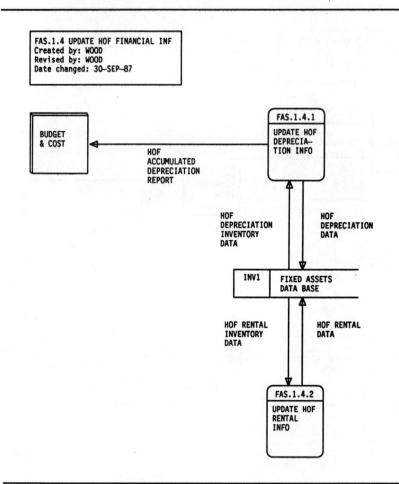

FAS.1.4 UPDATE HOF FINANCIAL INF
Created by: WOOD
Revised by: WOOD
Date changed: 30–SEP–87

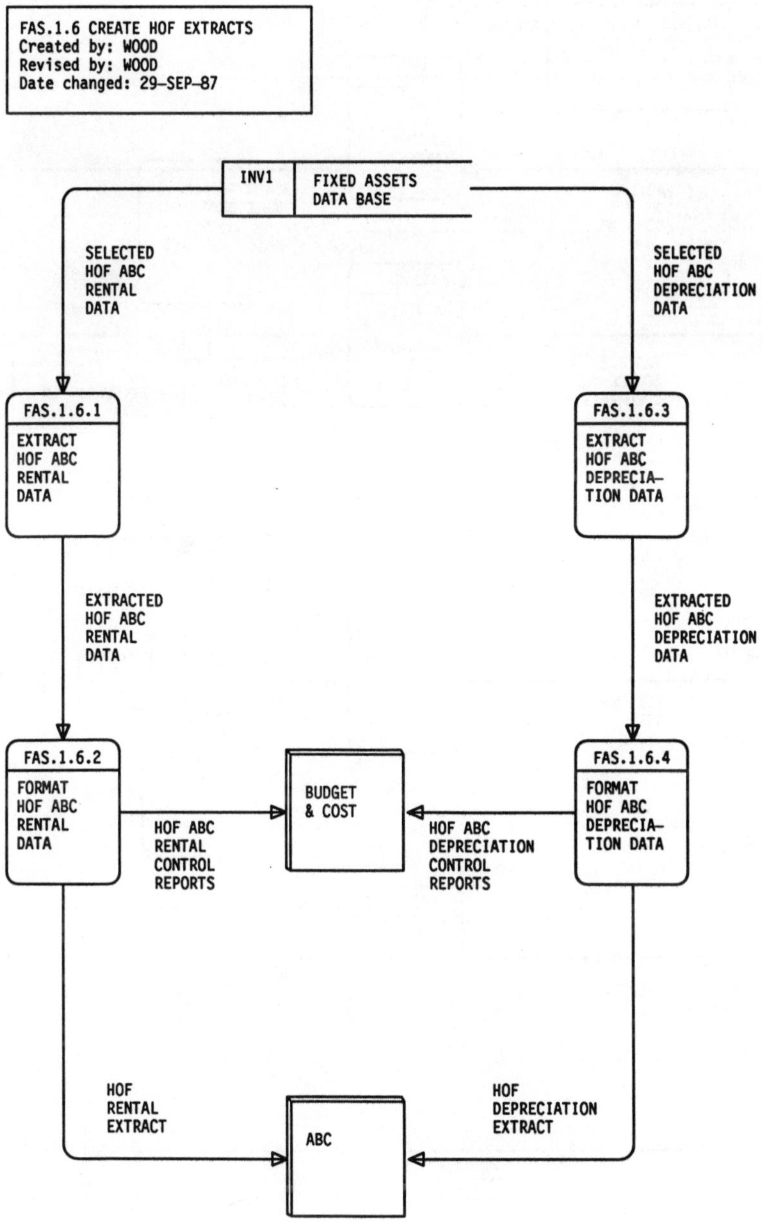

FAS.1.6 CREATE HOF EXTRACTS
Created by: WOOD
Revised by: WOOD
Date changed: 29—SEP—87

INV1 | FIXED ASSETS DATA BASE

SELECTED
HOF ABC
RENTAL
DATA

SELECTED
HOF ABC
DEPRECIATION
DATA

FAS.1.6.1

EXTRACT
HOF ABC
RENTAL
DATA

FAS.1.6.3

EXTRACT
HOF ABC
DEPRECIA—
TION DATA

EXTRACTED
HOF ABC
RENTAL
DATA

EXTRACTED
HOF ABC
DEPRECIATION
DATA

FAS.1.6.2

FORMAT
HOF ABC
RENTAL
DATA

BUDGET
& COST

FAS.1.6.4

FORMAT
HOF ABC
DEPRECIA—
TION DATA

HOF ABC
RENTAL
CONTROL
REPORTS

HOF ABC
DEPRECIATION
CONTROL
REPORTS

HOF
RENTAL
EXTRACT

ABC

HOF
DEPRECIATION
EXTRACT

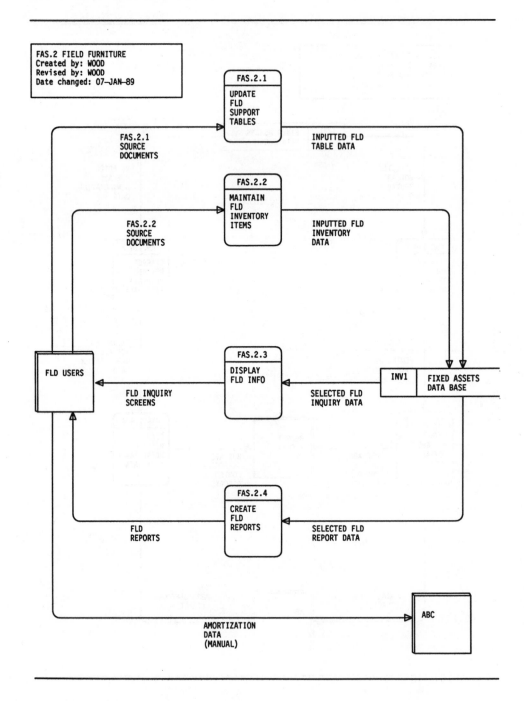

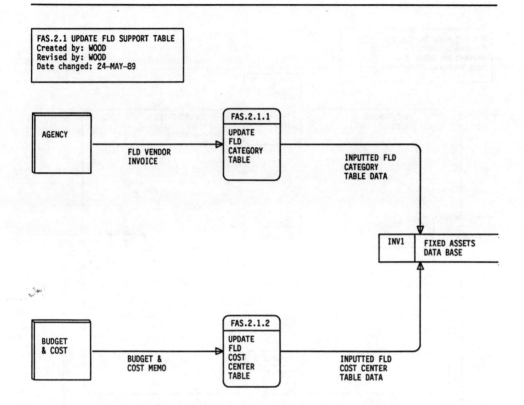

FAS.2.1 UPDATE FLD SUPPORT TABLE
Created by: WOOD
Revised by: WOOD
Date changed: 24-MAY-89

AGENCY

FLD VENDOR
INVOICE

FAS.2.1.1

UPDATE
FLD
CATEGORY
TABLE

INPUTTED FLD
CATEGORY
TABLE DATA

INV1   FIXED ASSETS
DATA BASE

BUDGET
& COST

BUDGET &
COST MEMO

FAS.2.1.2

UPDATE
FLD
COST
CENTER
TABLE

INPUTTED FLD
COST CENTER
TABLE DATA

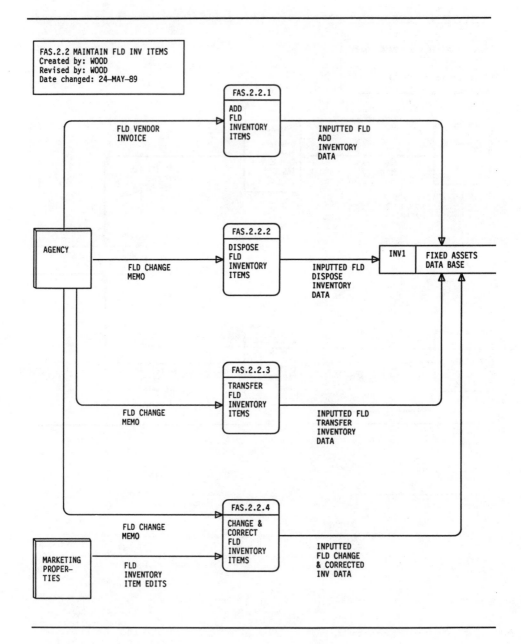

FAS.2.2 MAINTAIN FLD INV ITEMS
Created by: WOOD
Revised by: WOOD
Date changed: 24–MAY–89

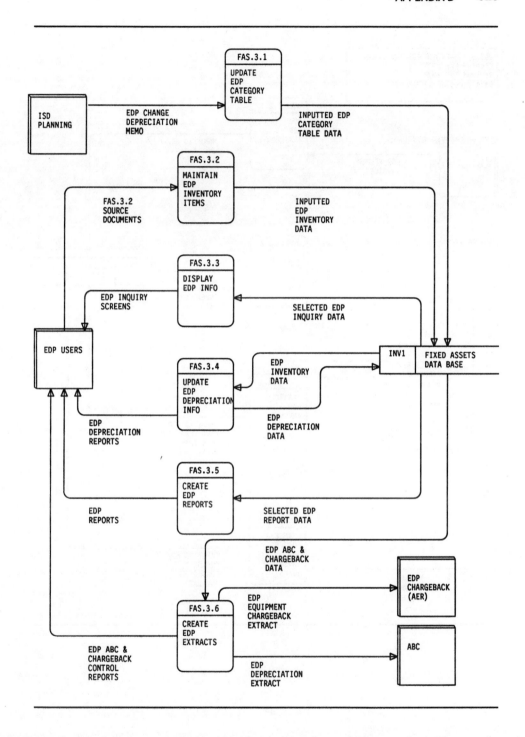

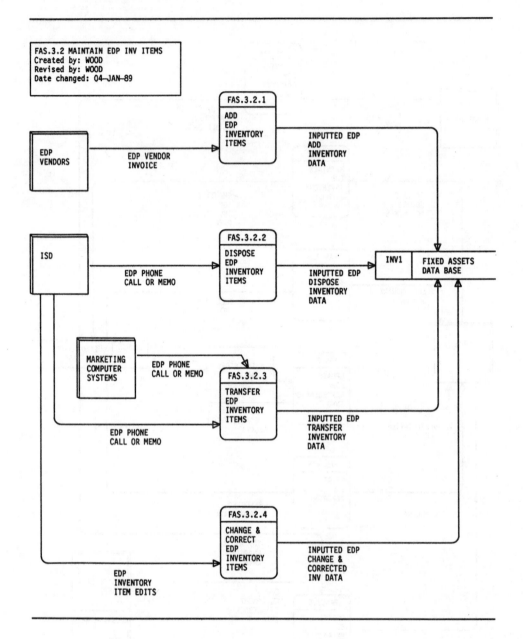

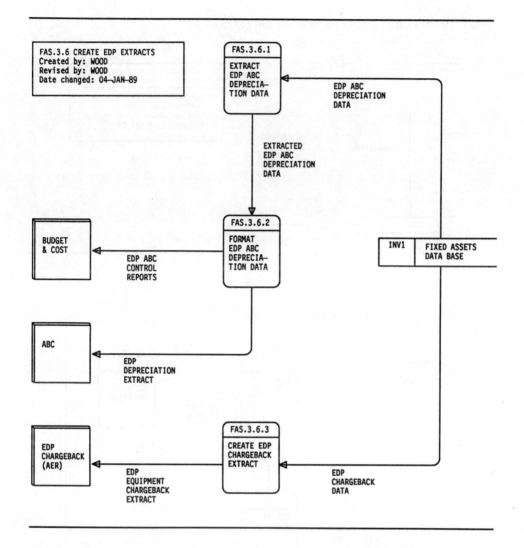

Interface with Cash Disbursement System (CDS)

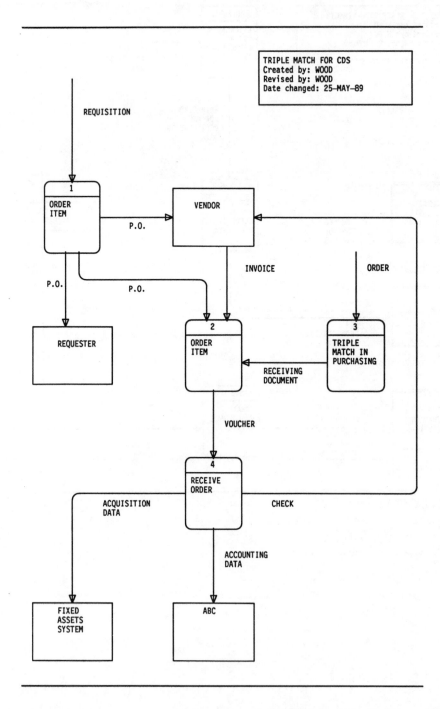

SCREENS

This section includes:

- a chart of the proposed screen flow
- a chart showing screen fields for each screen

Screen Flow

Figure 7 shows the proposed Fixed Assets screen flow.

Screen Descriptions

The chart on page 329 shows examples of which fields display on each screen.

PROCESSING

This section shows processing considerations for:

- calculations
- reversals
- partial updates

> *This section has been abbreviated to include only depreciation calculations and reversals.*

Calculations

The following defines the calculation methodology that will be used in the system and lists the actual calculations for:

- depreciation
- gain/loss
- monthly chargeback
- tax

Calculation Methodology

For statutory book depreciation, the following three methods will be hard-coded:

Fixed Assets Proposed Screen Flow

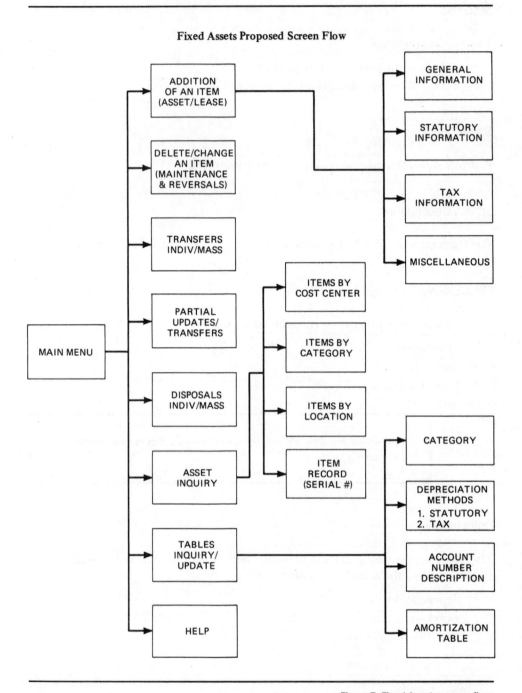

Figure 7 Fixed Assets screen flow

Screen Name (First Level)	Screen Name (Second Level)	Screen Fields	
Addition of Item	General Information	Asset type Company Asset Number Location Cost Center	Acquisition Date Purchase Price Voucher Number Quantity
	Statutory Information	Depreciation Period Depreciation Method	Depreciation Start Date Depreciation Status
	Tax Information	Depreciation Period Depreciation Method Depreciation Start Date	New/Used Federal Category Code
	Miscellaneous	Vendor Name Maintenance Amount Maintenance Effective Date	Maintenance Period Last Repair Date Repair Cost
Delete/Change an Item		Acquisition Date Purchase Price Depreciation Period Location	Cost Center Maintenance Amount Company
Transfers		Old Company Item Number Old Cost Center Old Location	New Company New Cost Center New Location Location (Mass Transfer)
Partial Updates and Transfers		Old Company Item Number Old Cost Center Old Location New Company New Cost Center New Location	Percent Transferred Amount Transferred Quantity Transferred Amount Added Percent Retired Amount Retired Quantity Retired
Disposals		Company Retirement Code Retirement Date Retirement Cost	Category Item Number Location (Mass Retirement)
Asset Inquiry	Items by Cost Center	Location Category Company	Purchase Date Purchase Price Chargeback Amount
	Items by Category	Cost Center Item Number Location Company	Purchase Date Purchase Price Chargeback Amount
	Items by Location	Cost Center Category Company	Purchase Date Purchase Price Chargeback Amount
	Item Record	Item Number Cost Center Location Company	Purchase Date Purchase Price Chargeback Amount
Tables Inquiry/ Update	Category	Description Chargeback Rate	Category ID Number Asset Type
	Depreciation Methods (Statutory)	Asset Type Depreciation Period Depreciation Method	Depreciation Percent Category ID Number
	Depreciation Methods (Tax)	Asset Type Depreciation Period Depreciation Method	Depreciation Percent Category ID Number
	Account Number Description	Asset Type Depreciation Expense Gain or Loss Accumulated Depreciation	Proceeds Account Asset Account Suspense Account

- straight line
- sum of the year's digits
- declining balance (where the user chooses from three different percentages)

A fourth table-driven method will handle an optional depreciation method where rates are added to the table.

For tax depreciation, 10 to 15 tables are required by the Tax Planning department. More tables will be needed as tax laws change.

Depreciation Calculations

> *In this sample document, depreciation calculations are shown for Home Office Furniture only.*

- *Additions.* Depreciation will be calculated (and the corresponding accounting entries made) as follows for all additions during the current month:

  DEPRECIATION = SYSTEM DATE (MONTH/YEAR) − PURCHASE DATE *
  ((PURCHASE PRICE − SALVAGE VALUE) / 144)

- *Reversal of disposals.* When a disposal is reversed, the following calculation should be made for adjusting depreciation:

  DEPRECIATION EXPENSE = (SYSTEM DATE − DISPOSAL DATE) * ((PUR-
  CHASE
  PRICE − SALVAGE VALUE) / 144)

- *Change in purchase date.* When a purchase date is changed, the following calculation will be made for adjusting depreciation:

  DEPRECIATION EXPENSE = (SYSTEM DATE − NEW PURCHASE DATE *
  ((PURCHASE PRICE − VALUE) / 144)

- *Change in purchase price.* When a purchase price is changed, the following calculation will be made for adjusting depreciation:

  DEPRECIATION EXPENSE = (SYSTEM DATE − PURCHASE DATE * ((NEW
  PURCHASE PRICE − SALVAGE VALUE) / 144)

Reversals

Reversal processing may be required when a field is changed or a transaction is reversed. The accounting entries resulting from reversals will be passed to the Accounting Budget and Cost (ABC) system. One possibility is to automate this interface to ABC. However, for expensed accounting entries where the original entry is in the prior year, there will be no automatic feed to ABC. Figure 8 shows

Possible Changes	Accounts Affected
Fields	
Purchase price	Chargeback Expense Depreciation Expense Accumulated Depreciation
Purchase date	Chargeback Expense Depreciation Expense
Disposal price	Gain/Loss
Disposal date	Gain/Loss Chargeback Expense Depreciation Expense Accumulated Depreciation
Disposal reason	Gain/Loss Chargeback Expense Depreciation Expense Accumulated Depreciation
Transfer date	Chargeback Expense
Depreciation method	Depreciation Expense Accumulated Depreciation
Category ID	Chargeback Expense
Transactions	
Transfer reversal	Chargeback Expense
Disposal reversal	Gain/Loss Chargeback Expense Depreciation Expense
Additional reversal (deletion)	Chargeback Expense Depreciation Expense

Figure 8   Reversal processing

the possible changes that may generate reversal accounting and which accounts are affected.

REPORTS

This section lists the reports required for:

- Home Office Furniture
- Field Furniture
- EDP Equipment
- Home Office Equipment
- Field Equipment
- Leasehold Improvements
- Tax Reports

The reports are divided into existing and new reports. Samples are included for all existing reports and for some new reports.

Home Office Furniture

Existing Reports

- Accumulated Depreciation
- Transferred to ABC—Depreciation Furniture
- Detail List by Category ID
- Transaction Listing—Home Office Furniture
- Interface Control reports

New Reports

- Monthly Activity Summary
- Separate error reports for each report

Field Furniture

Existing Reports

- Cost Center report
- Items Location report
- Categories within Cost Center

New Reports

- Field Furniture—Monthly Amortization for Month Ended mm/yy/dd
- Field Furniture—ABC Interface for Month Ended mm/dd/yy
- Transaction Listing—Field Furniture
- Interface Control reports
- Separate error reports for each report

EDP Equipment

Existing Reports

- List of EDP Equipment by Cost Center
- Chargeback Interface
- Accumulated Depreciation by Item
- EDP Equipment List by Category
- EDP Equipment by Location

New Reports

- Summary Chargeback Interface for Month 19nn (combination of Chargeback Interface and Items by Cost Center)
- Monthly Activity Summary
- Transaction Listing—EDP Equipment
- Separate error reports for each report

Home Office Equipment

New Reports

- Home Office Equipment ABC Interface
- Accumulated Depreciation
- Home Office Equipment List by Category
- Home Office Equipment by Cost Center
- Transaction Listing—Home Office Equipment
- Monthly Activity Summary
- Separate error reports for each report

Field Equipment

New Reports

- Field Equipment List by Category
- Field Equipment by Location
- Transaction Listing—Field Equipment
- Summary ABC Interface for Month 19nn (by Cost Center)
- Monthly Activity Summary
- Interface Control reports
- Separate error reports for each report

Leasehold Improvements

New Reports

- Transferred to ABC—Depreciation Leasehold Improvements
- Accumulated Depreciation
- Detail List by Category ID
- Monthly Activity Summary
- Transaction Listing—Leasehold Improvements

- Leasehold Improvements by Location
- Interface Control reports
- Separate error reports for each report

Tax Reports

The reports listed below should be provided for all assets tracked by the system.

New Reports

- Tax Depreciation reports—Under General Depreciation System
- Tax Depreciation reports—Under Alternative Minimum Tax
- Gain/Loss reports to File IRS Form 4797
- Dispositions
- Acquisitions
- Year-to-Date Summary
- Summary by Purchase Year
- Tax Projection for Depreciation
- Tax Projection for Acquisitions and Retirements

This sample document includes only one report sample, which is shown in Figure 9.

OPEN ISSUES

This section shows all issues that need to be addressed after the JAD session. This includes the person responsible for resolving the issue and the date it should be resolved by. If more than one person is assigned to an issue, an issue *coordinator* is designated to manage the resolution of the issue. When the issue is resolved, the coordinator sends a written copy of the resolution to the executive sponsor and all the participants.

Issue 1: Information Services Operations Requirements

Assigned to: Jane Grais
                Michael Marks (coordinator)
                Lenora McGrath

Resolve by: 11/6/87

ACCUMULATED DEPRECIATION
FOR HOME OFFICE FURNITURE AS OF AUGUST 1987

CAT. ID	DESCRIPTION	NO ITEMS YR	EOM	PURCHASE PRICE	ACCUM DEP EOM	NO DISP	COST DISP	DISP ACCUM DEPR	DISP PRICE	REGULAR DEPR
99999	Open shelving	99	999	99,999.99	99,999.99	999	99,999.99	99,999.99	99,999.99	99,999.99
99999	Cashier safe	99	999	99,999.99	99,999.99	999	99,999.99	99,999.99	99,999.99	99,999.99
99999	Lockers	99	999	99,999.99	99,999.99	999	99,999.99	99,999.99	99,999.99	99,999.99
99999	Bookcase, small	99	999	99,999.99	99,999.99	999	99,999.99	99,999.99	99,999.99	99,999.99
99999	Bookcase, large	99	999	99,999.99	99,999.99	999	99,999.99	99,999.99	99,999.99	99,999.99
99999	Letter holder	99	999	99,999.99	99,999.99	999	99,999.99	99,999.99	99,999.99	99,999.99
99999	Paper sorter	99	999	99,999.99	99,999.99	999	99,999.99	99,999.99	99,999.99	99,999.99
99999	Floor lamp	99	999	99,999.99	99,999.99	999	99,999.99	99,999.99	99,999.99	99,999.99
99999	Kutani jar lamp	99	999	99,999.99	99,999.99	999	99,999.99	99,999.99	99,999.99	99,999.99
99999	Handled urn lamp	99	999	99,999.99	99,999.99	999	99,999.99	99,999.99	99,999.99	99,999.99
99999	Wall whiteboard	99	999	99,999.99	99,999.99	999	99,999.99	99,999.99	99,999.99	99,999.99
99999	Desk chair	99	999	99,999.99	99,999.99	999	99,999.99	99,999.99	99,999.99	99,999.99
99999	Desk chair w/wheels	99	999	99,999.99	99,999.99	999	99,999.99	99,999.99	99,999.99	99,999.99
99999	Conference chair	99	999	99,999.99	99,999.99	999	99,999.99	99,999.99	99,999.99	99,999.99
99999	Arm chair w/wheels	99	999	99,999.99	99,999.99	999	99,999.99	99,999.99	99,999.99	99,999.99
99999	Tackboard	99	999	99,999.99	99,999.99	999	99,999.99	99,999.99	99,999.99	99,999.99
99999	Conference cabinet	99	999	99,999.99	99,999.99	999	99,999.99	99,999.99	99,999.99	99,999.99
99999	Laminate cabinet	99	999	99,999.99	99,999.99	999	99,999.99	99,999.99	99,999.99	99,999.99
99999	Walnut lectern	99	999	99,999.99	99,999.99	999	99,999.99	99,999.99	99,999.99	99,999.99
99999	Round table, small	99	999	99,999.99	99,999.99	999	99,999.99	99,999.99	99,999.99	99,999.99
99999	Round table, large	99	999	99,999.99	99,999.99	999	99,999.99	99,999.99	99,999.99	99,999.99
99999	Conference table	99	999	99,999.99	99,999.99	999	99,999.99	99,999.99	99,999.99	99,999.99
99999	Credenza	99	999	99,999.99	99,999.99	999	99,999.99	99,999.99	99,999.99	99,999.99

Figure 9 Accumulated Depreciation Report

Description: What are Information Services Operations (ISO)'s requirements for the Fixed Assets system? Are there any online real time requirements? If so, would ISO pay for these? Some possible inventory requirements include:

- items by desk location (employee name and telephone extension)
- terminal parts sent out for repair
- the diversity of field equipment such as 3B2s, Wangs, and HP Laser printers (this could be shown in a profile by agency)
- inventory and maintenance information via ad hoc reporting

This issue is resolved as described in the memo below.

November 10, 1987

To: Josephine Rose

From: Michael Marks

Subject: Resolution of Issue Number 1 (Information Services Operations Requirements)

In a meeting with Jane Grais, Lenora McGrath, and myself, we determined that:

- ISO has no need for online real time updating. Batch processing will suffice.
- ISO must be able to further locate and identify EDP items. Fields will be added to the data base to handle this.
- Input screens will be added to the system for ISO information updates. This information will be maintained by ISO.
- Inventory items will be accounted for by an inventory cost center and transferred to a new location cost center when an item is put in service.
- A major concern of ISO is the timely transfer of items between cost centers. They need quick turnaround. We will discuss this with the Finance department.
- Reports for ISO include:

  Weekly   Remote Terminal List (for the home office)
             Local Terminal List (for the data center)

  Monthly   Inventory by Location and Serial Number
             Installations and Removals by Cost Center

  Quarterly Remote Terminal List by Cost Center
             Local Terminal List by Cost Center

  On Request Inventory by Agency
             Inventory at Home Office by Item and Serial Number
             Response Time for Remote and Local Terminals

In summary, ISD can accommodate all ISO needs concerning the Fixed Asset system.

Issue 2: Capital and Operating Leases

Assigned to: Hugh Close
                Stella Kowalski (coordinator)

Resolve by: 11/6/87

Description: Will the Fixed Assets system support both capital and operating leases? (Capital leases are long-term leverage leases where at the end of the lease the asset is owned. Operating leases are for rentals only.) If the Fixed Assets system will support both types of leases, the following reports would be required:

- Detail Lease
- Lease Amortization Schedule
- Capital Lease Depreciation

This issue is resolved as described in the memo below.

November 12, 1987

To: Josephine Rose

From: Hugh Close
          Stella Kowalski

Subject: Resolution of Issue Number 2 (Capital and Operating Leases)

We determined that the company must safeguard all equipment used in its operations whether it is leased or owned outright. Therefore, the Fixed Assets system must account for equipment that is under lease. For equipment under capital leases, a "cost" will be assigned and entered into the system. Depreciation will be calculated just as for owned equipment. For operating leases, only the information required by ISO Facilities Management (such as location) should be tracked by the system.

Other accounting considerations associated with capital leases are not necessary for owned equipment.

We would like ISD to estimate these additional features separately. One software package offers an "add-on" to account for leased equipment under capital leases for an additional $8,000. The system must handle both book and tax treatments.

REFERENCES

Brown, Darlene. 1987. Everyone's talking about JAD. *GUIDE 69* (Atlanta, Georgia), November 2, Session no. MP5471A: 5–6.

Carroll, Lewis. 1967. *Through the looking glass.* New York: Collier Books. pp. 299–300.

Cosby, Barbara A. 1985. Planning for success: The importance of JAD pre-work. *GUIDE 61* (Anaheim, California), March 7, Session no. MP7436A: 7–8.

Gill, Allen. 1987. Setting up your own group design session. *Datamation*, November 15: 88, 92.

Godfrey, Laura E. 1986. Joint application design—A timesaver. *Resource*, March/April: 28.

Hennie, Dale P. 1985. Minding Ps and Qs brings systems gains. *Resource*, May/June: 44.

IBM. 1984. *JAD overview pamphlet.*

Martin, James. 1984. *Information systems manifesto.* Englewood Cliffs, NJ: Prentice Hall.

McClure, Carma. 1988. Implementation: Strategies for success. *System Builder*, October/November: 31.

New York Life. 1986. Joint application design. Paper presented at LOMA Systems Forum, Information Systems Committee, p. 1.

Parikh, Girish. 1982. *How to measure programmer productivity.* Wellesley, Massachusetts: Q.E.D. Information Sciences, Inc. pp. 10, 11, 12.

Petzold, Kent. 1987. CASM provides MIS with new return on software investment. *InformationWEEK*, April 27: 31.

Rush, Gary. 1985. A fast way to define system requirements. *Computerworld*, October 7: 11–12.

Western-Southern Life. 1986. Modified joint application design sessions. Paper presented at LOMA Systems Forum, Information Systems Committee, p. 4.

BIBLIOGRAPHY

Andrews, Dorine C., and Steven D. Rind. 1985. User-driven design: A way to computer creativity. *The Office*, May: 171–172, 229.

Benham, Barbara Tzivanis. 1985. Reducing confusion in project design. *Best's Review* (Life/Health Insurance Edition), September: 86–90.

Brown, Darlene. 1987. Everyone's talking about JAD. *GUIDE 69* (Atlanta, Georgia), November 2, Session no. MP5471A, 1–33.

Chamberlain, Wilt, and David Shaw. 1973. *Wilt*. New York: Macmillan Publishing Co., Inc. pp. 134–137.

Cosby, Barbara A. 1985. Planning for success: The importance of JAD pre-work. *GUIDE 61* (Anaheim, California), March 7, Session no. MP7436A, 1–10.

DeMarco, Tom. 1978. *Structured analysis and system specification*. Englewood Cliffs, NJ: Yourdon Press.

Doyle, Michael, and David Straus. 1976. *How to make meetings work*. New York: The Berkley Publishing Group. pp. 3–4.

Gane, Chris. 1987. *Rapid systems development*. New York: Rapid System Development, Inc.

Gill, Allen. 1987. Setting up your own group design session. *Datamation*, November 15, 88–92.

Godfrey, Laura E. 1986. Joint application design—A timesaver. *Resource*, March/April, 24–28.

Hennie, Dale P. 1985. Minding Ps and Qs brings systems gains. *Resource*, May/June, 43–46.

IBM. 1984. *JAD overview pamphlet*.

Leavitt, Don. 1987. Team techniques in system development. *Datamation*, November 15: 78–86.

Martin, James. 1984. *Information systems manifesto*. Englewood Cliffs, NJ: Prentice Hall.

McClure, Carma. 1988. Implementation: Strategies for success. *System Builder*, October/November: 31.

New York Life. 1986. Joint application design. Paper presented at LOMA Systems Forum, Information Systems Committee.

Parikh, Girish. 1982. *How to measure programmer productivity*. Wellesley, Massachusetts: Q.E.D. Information Sciences, Inc.

Petzold, Kent. 1987. CASM provides MIS with new return on software investment. *InformationWEEK*, April 27: 31.

Rush, Gary. 1985. A fast way to define system requirements. *Computerworld*, October 7: 11–16.

_____. 1985. Facilitated application specification techniques. *GUIDE 61* (Anaheim, California), March 7, Session no. MP7436B:1–5.

_____. 1984. JAD project aids design. *Computerworld*, December 24: 31, 38.

Watling, Richard J., Jr. 1987. Systems requirements workshops. *Proceedings of the Insurance Accounting and Systems Assn., Inc.*, Session no. 156: 269–271.

Weinberg, Victor. 1980. *Structured analysis*. New York: Yourdon Press.

Western-Southern Life. 1986. Modified joint application design sessions. Paper presented at LOMA Systems Forum, Information Systems Committee.

Wetherbe, James C. 1988. Traditional approaches to systems development. In *Systems Development Management*. Pennsauken, NJ: Auerbach Publishers Inc.

Yourdon, Ed. 1988. *Managing the systems life cycle*. Englewood Cliffs, NJ: Yourdon Press.

_____. 1985. *Structured walkthroughs*. Englewood Cliffs, NJ: Yourdon Press.

INDEX